THE
FUNDAMENTALS
OF PRODUCT
DESIGN

BLOOMSBURY VISUAL ARTS
Bloomsbury Publishing Plc
50 Bedford Square, London, WC1B 3DP, UK
1385 Broadway, New York, NY 10018, USA
29 Earlsfort Terrace, Dublin 2, Ireland

BLOOMSBURY, BLOOMSBURY VISUAL ARTS and the Diana logo
are trademarks of Bloomsbury Publishing Plc

First published in Great Britain 2009 by AVA Publishing
This edition published by Fairchild book 2016
Reprinted by Bloomsbury Visual Arts 2023

A catalogue record for this book is available from the British Library.

Library of Congress Cataloging-in-Publication Data
Names: Morris, Richard, 1962-author.
Title: The fundamentals of product design / Richard Morris.
Description: Second edition. | London; New York: Fairchild Books, an
imprint of Bloomsbury Publishing, Plc, [2016] |
Series: Fundamentals | Includes bibliographical references and index.
Identifiers: LCCN 2016004509 | ISBN 9781472578242 (pbk.: alk. paper) |
ISBN 9781472578259 (epdf: alk. paper)
Subjects: LCSH: Product design. | New products.
Classification: LCC TS171.M68 2016 | DDC 658.5/75–dc23
LC record available at http://lccn.loc.gov/2016004509

ISBN: PB: 978-1-3503-9885-6
ePDF: 978-1-4725-7825-9

Series: Fundamentals

Typeset by Lachina
Printed and bound in Great Britain

To find out more about our authors and books visit
www.bloomsbury.com and sign up for our newsletters.

THE
FUNDAMENTALS
OF PRODUCT
DESIGN

SECOND EDITION

RICHARD MORRIS

BLOOMSBURY VISUAL ARTS
LONDON • NEW YORK • OXFORD • NEW DELHI • SYDNEY

Contents

Introduction

People have been designing products for a long time. Some of the earliest known stone tools date back nearly two million years and it's likely that other artifacts made from less robust materials such as rope, leather, and wood were made long before this. Animals, too, have learned to design, with all sorts of mammals, insects, and invertebrates capable of creating beds and doorways or fashioning poking and hunting implements. This innate capacity to create things that help us through life might explain why the act of design is so intrinsically satisfying.

The Industrial Revolution changed the nature of design by moving the focus of

Herman Miller
Mirra chairs

creating artifacts for people mostly located around us to creating products on a massive scale for larger markets of unknown and distant peoples. When design thinkers at the turn of the twentieth century began to challenge the utilitarian nature of these early mass-produced goods, the first industrial designers such as Christopher Dresser and the Deutscher Werkbund began to appear. These were the specialists working to improve the appearance and functionality of ubiquitous goods such as furniture and kitchen appliances.

Designers still face the task of making goods for unknown consumers, which must work well and look good but also cope with additional challenges that were less prevalent to our design forbears. They must work within a massively complex world, gathering, processing, and synthesizing vast quantities of information within a dynamic and changing environment. They must also work quickly and accurately in what is a ferociously competitive world.

The fact that designers can cope with these complex, modern challenges is because they usually follow some kind of a process. In the United States, the term *industrial design* has evolved from its original roots of form and function to describe this broader design process. In Europe, the term *industrial design* has remained more focused on its original meaning of function and form, and the term *product design* has come to define the design process. This confusingly gives us two phrases that at once describe both different and the same things depending on where you come from.

Whether you prefer the term *product design* or *industrial design*, the aim of this book is the same, to bring clarity to the design process. This is a big challenge for one book because the process covers such a vast range of topics. By highlighting some of the fundamentals, however, it is hoped to help you the designer to better understand the process and to see some of the tools and techniques at work. It should help you consider your own style, ideas, and strengths and weaknesses in this beautiful and satisfying practice. In so doing, it aims to help you to create desirable, innovative, viable, and well-functioning products. It is through you that we can all enjoy a better world in which to live.

The design process

This book adopts the simplest of models to explain the design process, a model that represents design as a straightforward linear sequence. The sequence starts with research into a problem and then proceeds to generate concepts to solve that problem followed by the fleshing out of one concept through product development. The emerging product is then made real through production and finally launched into the market.

Research ⇨ Concepts ⇨ Development ⇨ Production ⇨ Launch

Note also that there are other models to describe the design process than the simple linear model, such as the double diamond, total design, the waterfall, or the Christmas tree model. The Christmas tree model, for example, suggests that each stage of the process might require much wider thinking at the outset before it begins to focus in so that the process begins to look like a Christmas tree.

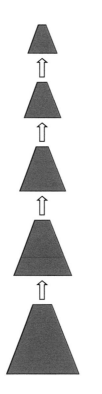

A straightforward linear model is used within this book, however, because it is simple to understand, and by reading through from start to finish you can easily gain a holistic view of the process. Along the way, you should also get to see the breadth and depth or work involved in design. Some people, including younger designers, sometimes think that design is just about good ideas and sketching, but this book should illustrate some of the other capabilities required. A designer is more likely to use an armory of everyday skills such as web searching, reading, visiting, meeting, filing, noting, experimenting, talking, planning, observing, and presenting as they are sketching and ideating. This is why the great inventor Thomas Edison described genius as "one percent inspiration and ninety -nine percent perspiration."

You do not need to read the book from start to finish and can dip into relevant sections as you work through a design project. The sections are organized to help you navigate in this way. The information provided may direct you toward areas that you need to think about, help you understand issues, or point you toward sources of further information.

You might even just elect to flick through the book at random, particularly useful if your own design work is getting tough. When you are struggling for time and beset by difficulties, select some pages and be inspired and motivated by the outcomes of good design being shown. Remember that the products and ideas you see probably did not appear without some effort and anxiety.

"Design is a plan for arranging elements in such a way as best to accomplish a particular purpose."

Charles Eames

Product research

Students of design are generally keen to get started on a design project straight away, to show how creative they are and to make their mark on the world. Sometimes they can be too hasty. Even experienced, professional designers can rush ahead too quickly, keen to make their own mark or concerned by over-anxious managers or the need to hit rapidly moving markets. A designer should, however, always start cautiously by questioning the purpose of the project itself:

- Is what is being asked clear?
- Is it viable (in other words, is it worth the effort)?
- Is it ethical—does it fit your own moral compass?
- Is it the right project?

It's essential to ask these types of questions because developing new products is usually expensive, with jobs, reputations, and company futures on the line if the project is wrong from the outset. Only once these questions are answered should the design process begin—and this doesn't mean now generating lots of ideas to show how creative you are. The final product is much more likely to be successful if the fullest possible understanding of the issues and requirements around the design challenge are understood. The first step therefore starts with Product Research. This might include liaising directly with users and consumers (primary research) or from more desk-bound exploration (secondary research). This chapter sets out some of the methods and tools for conducting primary and secondary research and the need to evaluate capturing this information to help define the requirements for the future product.

Figure 1.1
Morph device
concepts by Nokia   **1.1**

"Research is what I'm doing when I don't know what I'm doing."
Wernher von Braun

Designing for people

All products link in some way with people, and the first place to start researching is therefore to better understand the needs (and wants) of people. It might seem foolish to embark on a project without knowing as accurately as possible what people really need, but it is incredible how many designers or companies forget this, perhaps hoping that clever technology or slick marketing will sell a product that doesn't do its job in the first place. Good designers know how to design for others, not just for themselves or for some "blind" market. This aspect of putting people to the fore of the design process is referred to in terms such as Human Centered Design, human factors design, or user-centered design.

Research methods

You can't realistically ask all 7 billion people in the world what they need to solve a problem or what they think a product should have. Consequently, good product research is about the approaches and tools that can gather helpful information and insights in the most effective way.

This research should not only uncover the core requirements of a product, but should also aim to uncover the key or unique insights which might lead to a new, innovative, and perhaps world-changing product. Striving for this excellence can be a unique driver that helps to define product designers. User-centric companies, such as IKEA, Apple and IDEO,

"If you want to understand how a lion hunts don't go to the zoo. Go to the jungle."

Jim Stengel, former CMO, Procter & Gamble

WICKED PROBLEMS

"Wicked problems" are problems that might be less well understood—perhaps less obvious—and that have incomplete, contradictory, or shifting parameters making them difficult to find solutions for. These can be great challenges for designers and ideal ground for innovation.

also know the importance of this quest. These companies, which you probably recognize as being innovative, are likely to use a strategic rather than an ad hoc approach to researching the needs of people. Good Product Design then is as much about targeting, managing, and organizing market information on an ongoing basis (including recourse to "big data") as well as team work, communication, and change management as it is about individual creativity and visualization skills.

Figure 1.2
Personal environment monitor by Lapka.
Increas-ingly aging societies are opening up opportunities for medical and well-being devices. This monitor uses a choice of modules to sense the quality of the environment around you, placing this data in context and relaying the information to you through your phone directly or as part of an evolving story of your world. Materials include high-strength and scratch-resistant polyoxymethylene.

1.2

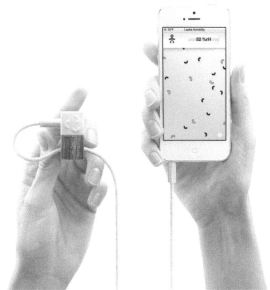

"I don't have any furniture of mine in my room."
Marc Newson

Primary research

Primary user research means gathering new and firsthand data directly from people. It is a form of empiricism (understanding through experience) but is based on evidence rather than intuition and is achieved through techniques such as interviews, surveys, questionnaires, and focus groups.

Statistically, it is not necessary to talk to every person at whom your product might be aimed. Somewhere between 100 and 1,000 responses is often cited as a representative spread of people; however, the more people you ask, the more likely your information will be accurate. Internet technologies have opened up opportunities for exploring a much wider range of opinions quickly and at low cost using formats such as e-mail shots, websites, forums, blogs, social networking sites, or gaming sites. The neologism *crowdsourcing* has been used to capture this method of working, although this is sometimes referred specifically to the use of paid, off-site helpers rather than the general public, which is then more correctly referred to as *opensourcing*.

Weaknesses in this form of direct research can arise from group think or leading questions, which can lead people to give false answers with disastrous effects on the resulting design process. Leading question might be "Do you like my idea?"—particularly if you have asked your family and friends for their opinions. There is also often some validity in the truism that people often don't actually do what they say they do (or do what they think they do).

Good technique can help to alleviate bias in the research. For example, iterating between "how" and "why" questioning to gain the emerging picture, clustering responses to find the route to the important parameters, forcing decisions rather than having "don't know" or "maybe" choices, and using "should it be smaller, larger or stay the same"-type questions can help the respondent.

"I wore a colostomy bag for a week and I've also bought prosthetic breasts and tried on a bra. I've stabbed myself with injection devices and I've dressed up in surgical gear."

Alun Wilcox at PDD

1.3

Empathic modeling

Empathic modeling or personal observation is an empirical research philosophy drawing from approaches used in anthropology (the study of humanity) and concerns using research in the field within cultures or societies to gain a personal "feel" for a problem. Within product design, this translates to the belief that product knowledge can be gleaned from the designer's own empathy and experience of the problem or situation being addressing. Designers can draw from their past experiences or from new experiences established through trial and error (heuristics) or situational testing such as role playing with specific products. This is like method acting within the film industry, where the actor lives and feels a part before filming in order to understand the emotions of the character. One of the beauties of a career in design is that this empathy can be achieved by anyone and does not need years of experience and in many cases—designing for younger people for example—might be best understood by a younger mind.

It is, however, important to recognize that although this technique is a rich area for understanding, on its own its results can be easily be tainted by our own prejudices and perceptual limitations. A designer might draw on his or her own experiences, but what if that view is not in line with what most other people want? People will either not buy your product or buy your competitors instead if it better fits their needs. Also, what if the designer has no experience to draw on—if the project was for the elderly, or children, or infirm for example?

Figure 1.3
Urban beehive by Philips. Many trends point toward the desire for a more organic future in response to environmental concerns and a human desire to reconnect with nature. This beehive concept plays with the idea of making the home a center for biosystems and is part of a "probing" activity designed to explore future ideas and stimulate discussion.

Ethnography

Understanding and insight can be obtained through the anthropological technique of **ethnography**, which involves observing and interpreting behaviors. In design, this might mean gaining insights by watching how people behave in certain scenarios or interact with products or prototypes. This can be done in situ for example in homes or places of work, or as part of a specific test. Behind the obvious activities may be deep meanings of behavior or cognitive human factors, (such as understanding how people absorb information). Observations are, therefore, usefully recorded to allow events to be watched repeatedly for the detailed and sublime clues and messages that may be involved enabling a **cognitive model** to be constructed.

Anthropological methods such as ethnography can open up issues of ethics and validity. For example, people who are being video-recorded may change their behavior simply because they are being recorded. Results may also be open to differing interpretations depending on the perspective of the observer/designer. If observation is done more discretely, then there are ethical issues around observing people without their knowledge.

1.4

Figure 1.4
Merck Serono easypod® design by PDD Group Ltd. A product designed to make hormone injections easier, principally for younger people. The device aims to improve confidence, control, and adherence to treatment regimes.

CULTURAL PROBES

Cultural probes is the name given to a technique used to gather information about the values and thoughts of people directly rather than through interpretation by others. It involves introducing people to evocative tasks and artifacts and allowing them to record their own feelings and responses using diaries, cameras, or written notes. The aim is to use a more creative approach to revealing insights and inspirations rather than the more formal and scientific techniques used in other forms of product research.

Understanding

Understanding the motivations, perceptions, beliefs, and attitudes of people can be illuminating and instructional but difficult to perceive and comprehend without some supporting theoretical frameworks. Psychology is the analytic and scientific study of mental processes and behavior and includes concepts such as perception, cognition, emotion, personality, behavior, and interpersonal relationships. By attempting to understand the roles that these functions play, a designer may be able to design better products. Like observation, this can produce powerful, insightful information that people might not recognize or be able to articulate in a direct interview. For example, the design of a toy for a three-year-old child might be safer, or more engaging and educational, if there is a greater understanding of a child's general cognitive and behavioral functions at that age.

Sociology is the understanding of human needs in a working world context through social or cultural interpretations. In this case, products are viewed very much as "artifacts," having meaning and relevance to the world around them beyond pure function. A study of the semantics (meaning) and development in time of a product, or related products, might help to generate a better understanding of the products underlying needs and requirements.

1.5

Figure 1.5
Switch screwdriver by Mr. Yunlong Liu, Mr. Peng Jia, Mr. Peng Cheng, Mr. Dongdong Wang, and Miss Yaoyao Xin. Exploring problems can sometimes release simple "Why didn't we think of that before?" solutions such as this screwdriver, which can be adapted to provide more torque for stubborn screws.

1.6

Figure 1.6
Sun Jar by Tobias Wong for Suck UK. Seasonal
affective disorder (SAD) is a type of depression that
can be triggered by the shortened daylight hours of
winter. Light therapy can be an effective treatment
for SAD. Although not a therapy itself, the Sun Jar
does effect a psychologically warming experience.
The jar collects sunlight via a solar collector, and
a battery and LED light also provide a practical
source of light at night.

"I roamed the countryside searching for answers to things I did not
understand."

Leonardo Da Vinci

Figure 1.7
Nike Air Jordan XX3. The Nike brand is often associated with both innovation and winning. Cool products can easily become outdated, however, so companies will often look toward new trends arising from young, dynamic social groups whose behaviors or symbols are likely to become the next benchmark of cool. "XX3" refers to the fact that this is Nike's twenty-third Air Jordan development.

Cool hunting

As the social groupings and subcultures of younger people become increasingly fragmented, it becomes harder to assess them with traditional research techniques such as questionnaires. It can also be difficult to assess a subculture if you do not understand it.

Cool hunting describes a combination of anthropological approaches, such as living with subgroups and observing with an understanding of the sociology and psychology at work. Researchers should be outgoing, have a wide network of contacts, and be capable of "deep hanging out." They should watch what young people are doing, where they are looking, what they are browsing, what badges they are wearing, how they are absorbing information, what communication methods are being used, and what clothes and attitudes they display and then interpret where these fads might go. Trends among young groups are important as they have a habit of eventually turning mainstream.

Figure 1.8
Nike Flyknit trainers. Flyknit trainers use single-strand construction to reduce weight and waste and are engineered to be strong at the precise points needed.

Secondary research

Secondary research involves exploring relevant information that is already available such as that gleaned from marketing reports or trade directories. This form of searching can be extensive—stretching to exploring a product's historic trends or including related products—and might include political and economic considerations. Fundamentally, just how big is the market and how much of that market could we hope to capture? This knowledge will help you understand from the outset if there is a cogent business argument to proceed with product development. The knowledge is also needed more downstream in the development process. For example, the size of the market can determine the manufacturing process and, therefore, some of the later design detail.

Secondary research also includes examining existing products to see how other designers are interpreting people's specific needs and problems. Who are they aiming their products at? What trends are there in the design? For example, examine the ranges, costs, materials, shapes, manufacturing processes, and so on. Explore which ones are more successful, and try to find out why they are more successful. Investigations should, therefore, examine its full range of competitor attributes including its features, signifiers, and benefits. **Reverse engineering** is the name given to the process of taking a product apart in order to understand how it works. Laser scanning makes this practice much easier to perform and is generally permissible as a research technique—but copying is not.

1.9

Figure 1.9
Neatly nesting mobile phones by Kyle Bean. Some trends become more obvious over time making it good to look back at past product ideas. The tendency toward smaller phones is illustrated by this set of nesting cardboard phone models.

Case study
Tord Boontje

Tord Boontje studied at the Eindhoven Design Academy and the Royal College of Art (where he later returned as Professor and Head of Design Products). Studio Tord Boontje was established in 1996 and has been based in both France and London. Boontje has a keen interest in social issues realized through projects in South America, and the quality of his work is recognized through many collaborations, awards, and showcased public collections.

Some companies invest heavily in facilities that will help them to fully understand consumer needs. For example, Gillette's U.K. research base for example employs hundreds of people and uses a vast range of techniques including high-speed and hidden filming of batches of men shaving every day to explore issues such as blade angles, blade pressures, hand grips, arm movements, and shave patterns. In the United States, the Fisher Price Play Lab has used over 200,000 children to help develop around 60,000 toys using observational techniques within a monitored play space. At the other end of the scale, it can be down to the skill and insights of a single designer or a small team to provide the ideas for new products, drawing on their experiences and creativity to hit the winning formula.

Dutch-born designer Tord Boontje is an example of this type of small but insightful and creative designer. He and the Boontje Studio are renowned for experimental work across a range of clients and have regularly produced award-winning designs.

The Garland light shade, for example, is made from laser-cut and -etched metal and hits many of the needs of the consumer. The ability of the consumer to wrap the shade around a light bulb in many configurations latches onto the market trend needs for individuality and customization. Its playful shapes produce patterns of light akin to sunlight through leaves of a tree, hitting consumer targets for nature and organic shapes. The Happy Ever After instillation that followed provided an impactful setting of color, texture, and form, reacting to—or helping to set—a trend for design that brings emotional and sensorial content to people.

The Boontje Studio questions and challenges many design axioms: that modernist must mean minimalist, that contemporary must conform to modernism at the expense of historical trends, that technology is a good thing in its own right, even if it's devoid of a connection with people. More recent work has advanced social projects and thoughts about design as collectible art, which often includes thoughts around paradoxes and perceptions.

It is this type of exploration and questioning, fostered through traveling, relocations, and collaborations, that has helped Boontje generate ideas that lead to instant consumer demand. It is a design skill that he has extended into lighting, furniture, clothing, and interior design with demand from clients that include Yamaha, Hewlett Packard, Swarovski, and Moroso.

1.10

**Figure 1.10
Witches kitchen cook-
ware.** Kitchen design
trends tend toward
the smooth and plain
using colored glass.
The Tord Boontje studio
targets and redefines
this trend by reimagin-
ing the whole kitchen
experience. Leaves
pressed into the clay by
South American artisan
communities leave a
fossil-like pattern after
firing.

Targeting

It's unlikely that one design is going to meet the individual needs of every person on the planet Research should start to identify who the most likely consumers are, if people's needs are uniform, and, if not, how they vary among similar people, or among different ages, genders, or cultures. Therefore, crucial decisions have to be made as to who should and should not be the focus of the design project. This does not mean forgetting the wider market and hoping that you don't make sales there, but it does mean not necessarily undertaking the impossible task of trying to please everybody. Sometimes this can also mean breaking away from the ways that markets are traditionally classified.

Figure 1.11
Brunton Hydrogen Reactor™. Most people might articulate concerns over cost and environmental implications for charging appliances, and some will form a specific targetable market. Brunton has responded to this concern with a hydrogen cell recharged from water.

1.11

Product requirements

In addition to user needs and wants, products need to conform to a range of general or specific performance and safety requirements. Because these need to be ascertained from the outset, they must be understood and considered in the early stages of product research.

Legislation

Most countries have some form of legislative requirements—consumer law—to ensure that products are free from defects, including a defined notion of a "safe" product. The European Directive 1985 on Product Liability, for example, has been implemented across Europe, such as in the United Kingdom's Consumer Protection Act (1987). In the United States, the Consumer Product Safety Act applies. These laws can include making it illegal to sell a product that is intrinsically dangerous, and this may be enforced by government agencies with criminal penalties for failure. In the United States, the Consumer Product Safety Commission acts as the overseer of product safety (in some cases it acts such alongside other agencies such as the Department of Transportation or the Federal Trade Commission). In the United Kingdom, this overseer might be the Trading Standards Office or the Competition and Markets Authority. There are equivalent overseers in most countries. Governmental bodies may also have their own set of additional legal requirements for specific products. "New Approach Directives" in the European Union are for example aimed at safety and performance requirements for products such as toys, sports, and safety equipment.

Most products must, therefore, include a safety requirement, and research should identify the minimum requirements needed to deem a product safe and the levels to which the product might go beyond these.

Most countries also have "remedy" legislation, providing a civil law route for consumers who buy defective products. Products must also, therefore, have a fit-for-purpose requirement. Each country might also have legislation relating to costs, materials used, environmental protection, methods of fixing, labeling, or notifications of goods for export. All of these legislative requirements must be noted as part of the research into product requirements.

"Safety doesn't happen by accident."

Anonymous

Figure 1.12
Figure 1.12
LotusSport
Pursuit Bicycle
by Mike Burrows.
Mike Burrows
radical design
of a carbon-fibre
molded mono-
coque bike frame
broke away from
the steel tube
diamond-shaped
frame that had
been thought to
be pinnacle of
excellence but
was deemed 'ille-
gal' until cycling
rules changed.

1.12

Regulations

Rules and guidelines for certain products may be set by trade organizations, umbrella groups, or controlling authorities. One example of this would be the minimum performance requirements set by the **International Organization for Standardization (ISO)**. ISO, a nongovernmental body to which most countries are affiliated, sets industrial and commercial standards. Through treaties or interpretation by national bodies, many ISO standards will become statutory requirements. Some of the ubiquitous ISO standards that designers will commonly engage with include the 14000 series on environmental management systems and the 9000 on quality management systems. In addition, most countries will have their own national standards authority that will interpret ISO standards, or they may develop their own complimentary or alternative standards. The British Standard BS888, for example, is the British regulation that defines the ISO standard set on drawing and technical product documentation, and BS7000 is a guide to systems for managing the design process. The U.S. engineering drawing equivalents might be found in the Y14.100 series of the American Society of Mechanical Engineers.

Regulations should always be considered rather than taken as read. Some designers assume that if a design meets the performance levels set by regulations and guidelines, then it might automatically be considered a safe product. This is untrue—if the guide is inadequate to ensure safe working, then it follows the product will be unsafe, too. Innovative new products and technologies can also push the boundaries faster than administrative bodies can work. For example, this is particularly evident in sport or IT industries where the introduction of new ideas can challenge and overwhelm existing rules and regulations.

"Rules are for the obedience of fools and the guidance of wise men."

Douglas Bader

Safety

Designers generally have a "duty of care" to look after the people who interact with their products, which can be an underlying legal requirement as in the United Kingdom's principles of common law and tort of negligence or a more specific act of legislation. In the United States, this legislation can change from state to state but will generally revolve around principles of foreseeability and the nature and degree of factors involved.

Product research should then not just detail user and functional requirements but also detail safety requirements such as maximum weights, reach, light levels, and so on. When considering safety factors, it should always be remembered that some people will always use products in ways for which they were never

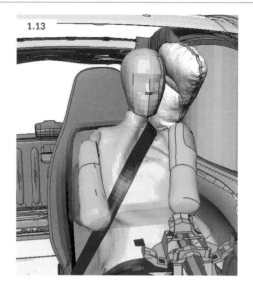

1.13

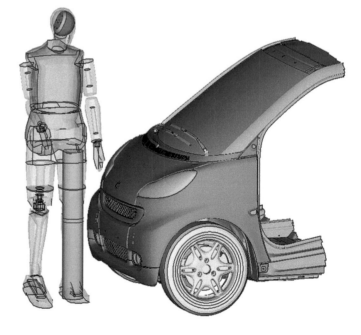

Figure 1.13
Smart Car. The cubic design of the Smart Car could make it vulnerable to impact damage in a collision. To overcome this, the car has a hemispherical steel shell that provides safety for occupants and also forms the car's chassis

intended. A furniture designer should always bear in mind that people always seem to be drawn to sit on a table, even though it is not designed for this purpose.

It should also be understood that the actual user may not be the target user the designer had in mind in the first place. The target user for a kettle could be a young, fit adult, but the person who actually picks up and interacts with the product might be a young child, or an elderly or infirm user, all with different levels of cognitive and physical abilities.

Ethics

There are normally ethical reasons for not wishing to be responsible for selling people defective products or worse still for causing injury to people. There may, too, be ethical reasons for not wanting to develop certain types of products. If you are environmentally minded, are you happy developing a throwaway product? If you are a pacifist, would you be happy developing a product for an industrial application that could ultimately have military benefits? Designers are professionals with a big impact on the world, and it's important to have an opinion. Sometimes designers have to weigh the balance between acting on principle and providing for themselves and family.

"A common mistake that people make when trying to design something completely fool proof is to underestimate the ingenuity of complete fools."

Douglas Adams

Technical research

At this early stage in the research process, it's useful to research types and trends in technologies that might help to generate a solution. This is not specifically to specify a type of technology that will be used but should instead gauge the levels of technical performance that might be achievable and the types of emerging technologies that might be useful (called **technology road mapping**).

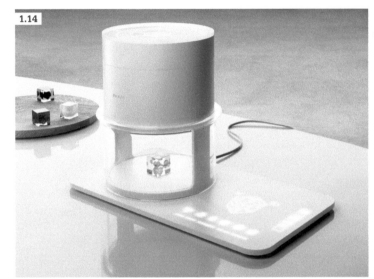

Figure 1.14
Philips food printer concept.
Technical complexity of ideas can often be appraised by drawing on the technical achievements in other product fields. The viability of this conceptual food printer could be appraised through an analysis of developing 3D prototyping techniques

Case study
Sir Jonathan Ive

Born in 1967, Jonathan Ive studied art and design at what is now Northumbria University. At 22, he became a partner in London-based design consultancy Tangerine, moving three years later to join the Apple design team in California, where he has since become Senior Vice President of Design. The Apple team has released the iconic iMac, iPod, iPhone, and Apple Watch. Ive's design work has been recognized by numerous awards and accolades including a British knighthood.

The research behind Apple's product design specification for the original iPod was to create a music device capable of storing up to 4,000 songs that could be accessed via a simple and intuitive interface. Behind these simple requirements, however, lay a host of additional technical and legal necessities relating to, for example, trademarks, copyright, software, file storage, data transfer, interfacing, connectivity, hardware, sound characteristics, and power. The iPod met these and surpassed many other user expectations by including features such as the iTunes downloading facility and by giving the iPod an iconic feel, created by a simple construction, using twin-shot polycarbonate-ABS snapped to a polished and laser-etched stainless steel back. The classic iPod also spawned a range of associated products, including the Touch, Nano, and Shuffle, which meet a range of targeted consumer groups, and each of these versions has been continually improved through a series of incremental developments. That Apple got the design and market details right has been demonstrated by hugely favorable industry and consumer reviews that have kept it at the head of a highly competitive market sector.

Having set the benchmark high with the iPod, Apple needed to again meet expectations with its mobile phone product. The original iPhone was small, easy to use, and solid, offering features that included a camera, portable media player (equivalent to the iPod), text-messaging functionality, visual voice mail, e-mail, web browsing, local Wi-Fi connectivity, and, of course, a flush multi-touch screen with a virtually rendered keyboard. Like the iPod, the iPhone has had to satisfy vast technical and legal requirements in addition to consumer requirements. Products must meet performance and safety standards in audio, visual, and power functions. To meet environmental needs, for example, the iPhone restricts PVC, bromine. and mercury in its production. It has also had to innovate continually to stay ahead of its competitors with the iPhone 6 coming in two different sizes to meet different user needs. These more recent releases have dispensed with the traditional "faux" materials and 3D shadow effects (effects referred to as *skeumorphism*) in favor of simpler and cleaner interfaces but avoided dullness by using layers of flat planes. Over the years, the iPhones have achieved outstanding sales figures, over 500 million units worldwide, achieving proportionally higher profits than its market share as a consequence of its embedded design excellence.

Jonathan Ive is quick to point out that the success of these products can be attributed to teams working together, which requires good communication and clear goals. He also cites "fanatical care beyond the obvious stuff" achieved through attention to detail in tools, materials, and manufacturing. Communications, goals, regulations, speed, knowledge, and assessment are not just the designer's responsibilities, however, and design-focused companies such as Apple are successful at embedding these issues into company culture as part of the core corporate strategy.

Figure 1.15a
iPhone by Jonathan Ive and the Apple design team.
The original iPhone combined three products in one: a phone, a media player, and an Internet device.
Photograph: Apple, Inc.

Figure 1.15b
iMac with Retina 5k display by Jonathan Ive and the Apple design team. Apple continues to lead the way in design excellence.

Defining needs

Design can be a breathtakingly wide activity, and although this can be invigorating and exciting, it can be challenging, too, requiring the simultaneous handling of complex requirements and information. Research in particular can produce a wide variety of data, and it's critical not to just interpret this into one long wish list. It's the skill of the designer to synthesize and translate this data into a cogently defined set of needs, ejecting "noise" (meaning unwanted information) and focusing on the key requirements that define the problem and provide signposts towards a future design solution.

Criteria

The descriptors that define a product's requirements, such as functions and cost targets, are known as criteria. It is possible to keep these in a designer's head, but it is preferable to write them down as an aid to clarity and as a way of communicating the product requirements to a larger audience. It can be common to find these criteria written as rather loose descriptions. For example, "the product should be big." In reality this is not helpful. "Big" to one person may be small to another, and how would you decide whether your finished product is big enough? It is, therefore, often helpful to define criteria precisely using where possible quantifiable metrics.

Figure 1.16
BeoLab loud speakers by Bang & Olufsen. Some companies might consider technical excellence as their priority market criteria, whereas others might select innovation or style. Bang & Olufsen attempts to satisfy all of these criteria across its range of speakers and electrical products and is largely successful in doing so. The BeoLab loudspeaker is enormously powerful and, with its circular cone design, can be placed anywhere in a room.

"The public don't know what they want; it's my job to tell them."

Alec Issigonis

Priorities

Some criteria will be more important than others; often criteria will conflict with each other. For example, if a product needs to be lightweight and durable, then there may come a point where making a design lighter begins to reduce its durability and hence shortens its lifetime. To help resolve these conundrums, it can be helpful to clarify which criteria have precedence. By prioritizing, designers can put the right emphasis into the right area and have some clarity when difficult decisions have to be made, which is particularly useful when groups of people are involved in a project.

Market criteria and priorities can have a number of related names and descriptions, for example:

- Market qualifying criteria are those attributes that a customer might expect from all similar competing products.
- Order-losing criteria are those considered to be critical to the customer and a failure to have them will exclude the supplier from consideration.

- **Order-winning** criteria are those attributes that enable a company's products to gain an advantage over their competitors in the eyes of the customers.
- **Unique selling points (USPs)** relate to the key feature or features that a potential product is likely to have to establish its "position" within the market.
- The Kano model, developed by Japanese theorist Professor Noriaki Kano, considered that criteria could be categorized into domains, often cited as expected features (basic), features that increase customer satisfaction as they improve (one-dimensional), and nonexpected features that differentiate the product (exciters).

PRIORITIZING CRITERIA

Design students will often put features of appearance or performance at the top of their criteria priority lists usually because this is a projection of their own desires or their course restrictions. Design professionals working in the real world will usually list safety and cost as the foremost requirements.

"Design is the most immediate, the most explicit way of defining what products become in people's minds."

Jonathan Ive

Product range

Another decision that needs to be made early is whether one product will meet the market need or whether a range of products is required. A straightforward market with uniform needs may be satisfied with one product. A complex market, with wide-ranging tastes, cultures, and requirements may require a range of products to be generated. This might require simple choices about color, to more complex choices about size, performance, and the modularity of a design. The iPhone 6 and iPhone 6 Plus are examples of this modularity: common technical platforms available in different sizes with different features and memory capacities available to meet different user needs. After-market cases and housings also allow users to individually personalize their products.

1.17

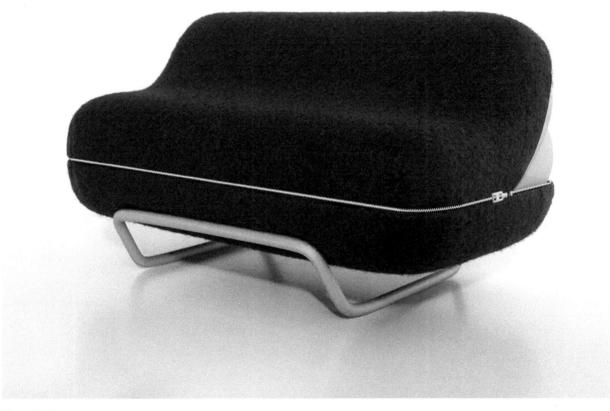

Figure 1.17
Eclosion by Olivier Grégoire. The Eclosion sofa is an example of a product that can be adapted to meet a wide range of different requirements, in this case inflating and adjusting the zip feature to fit individual comfort needs.

Benchmarking

Sometimes it's easy to get caught up in the excitement of a new idea and a potential new development. **Benchmarking** is the name given to the process of comparing and evaluating your position against that of others, and this can bring a sense of reality to the design challenge ahead. Drawing on the reverse engineering or evaluation of your competitor products can be useful in appraising the technical levels and standards that an industry is working to and that you need to meet or better.

1.18

Figure 1.18
Leatherman® Skeletool. For many years, Swiss army knives have been the acknowledged benchmark against which other penknives might be measured. The Leatherman® series has introduced new materials, tools, and ideas in a way that has redefined the benchmark. No longer a knife with accessory tools, the benchmark might now be considered as a multitool that includes blades.

Feasibility

An issue for designers is to know whether a new and innovative product will arrive at just the right time, or whether it will be too far ahead of its time for consumers. Similarly, a product conceived with advanced technologies might be achievable now, achievable in the very near future, achievable in the distant future, or just a distant dream. The rate of change in technology makes it very difficult to predict just how far to set the criteria, and sometimes difficult decisions have to be made whether to use known and existing technology or to risk trying something new.

The product brief

One of the key factors behind successful innovation is the ability to accurately capture and describe a problem and the research into it in a clear and accurate way, and this is usually contained with the **product brief** (or brief, and sometimes also referred to as a user requirement specification). The brief then outlines the requirements a future product needs in order to solve the problem and hence dictates the way in which the design will develop.

1.19

There is real skill in weaving together the key facts to create the most feasible summative picture. For some, this is an art form based on an intuitive feel for what's new and what's right; there is little or no reason to write anything down. For others, this will mean a lot more objective analysis based on a whole range of data including market, economic and political factors. Either way, the brief should contain enough information to define the concept but not constrain the subsequent development. It should include prioritized criteria, initial scope and constraints, and possibly some exemplar notional concepts. Words should be chosen carefully, as they can sometimes cause conflicts when meanings are misinterpreted.

Sometimes it can be hard to define a need through language, which can seem clumsy or overpowering. Trying to explain a color such as orange or a feeling of fear, using just words to convey just the right impact, can be very hard. Mood boards can be used as an alternative, using images, texts, samples, and colors in a composition that effectively demonstrates meanings and feelings without the contortions that words can sometimes cause.

Figure 1.19
Smart Crossblade. The concept and technology for the Smart Car is clear: it's a small urban car. It's not, however, sensible simply to make a big car smaller so the specification for this type of car requires attention to issues of comfort, space and economy. To meet these specifications, the Smart Car is lightweight and has almost cubic dimensions that give it the extra internal space and stability.

Constructing a consumer profile for a typical member (or members) of your chosen market can also be helpful. This should include not just demographic data such as age and gender but typical behavioral characteristics. It is the depth of knowledge behind these simple caricature presentations that makes this a very powerful technique.

Co-creation

Co-creation is an emerging process of partnership between designer and consumer where the designer works less in interpreting consumer needs and more in empowering consumers to develop their own design ideas. Its beauty is in avoiding any misinterpretation

Figure 1.20
Plume fountain pen by Vivien Muller. The brief for a pen might include cost, weight, durability, style, performance, and a host of other criteria. This plume/feather design pen places an elegant connection with the past as a primary design feature.

by the designer so that consumers get exactly what they want, but it requires a very different design perspective with systems and technologies that allow nontechnical members of the public to create the things they need. Fab labs and designer maker movements follow a similar ethos by empowering people to be able to design products for themselves.

Figure 1.21
Nissan IDx NISMO®. Many car companies use a common car architecture to allow different models to be produced efficiently. Nissan has gone one step further by using co-creation techniques alongside a common framework to allow consumers themselves to steer the development of different models—such as this IDx NISMO concept. If this variability can be done for cars, it can be achieved for much simpler products.

Case study
Thomas Heatherwick

Thomas Heatherwick has earned a reputation as an "ideas engine." He studied 3D design at the United Kingdom's Manchester Polytechnic and at the Royal College of Art. He was encouraged in his early years by the mentoring support of the designer Terence Conran and in 1994 established the Heatherwick Studio, which has since won many awards for its high-profile design work across the globe.

Design consultants really understand the need to put people and experience at the heart of a new product. They really understand the design process, and they really understand the need to innovate. They are also equipped with the tools and techniques needed to achieve this. This is why many companies turn to them for help when they need to launch a new product.

Thomas Heatherwick CBE, and the Heatherwick Studio of around 140 designers and makers, has met some challenging design briefs including the U.K. trade pavilion

at the Shanghai World Expo 2012 and the Olympic Cauldron for the London Olympics in 2012.

The studio tackled the brief for a new bus for London to replace the older Routemaster buses. Even though these older designs were iconic and a much-loved symbol of London, they were considered by many to be dangerous and expensive to build and operate. The new design would seek to capture this symbolic glory but allow for an increased number of stairs, improve disabled accessibility, and achieve a 40 percent reduction in the use

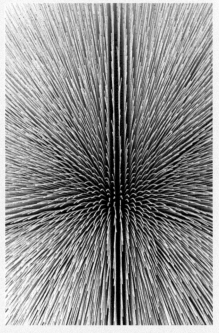

Figure 1.22
Seed Cathedral, U.K. Pavilion at Shanghai Expo 2010. The needs of the United Kingdom's trade stand at the 2010 World Expo included a requirement to "stand out" and to project a modern view of the nation using a limited budget. The award-winning design achieved this by using simplicity, clarity, and texture above technology and architectural complexity.

1.23

Figure 1.23
Spun Chair. A collaborative and experimental approach to design has generated the innovative Spun Chair design.

of fossil fuels. The Studio added to this brief the requirements for better driver visibility and a calmer traveling experience for passengers. The solution included removing internal clutter, providing the shell with ribbons of glass to give extra light, and, unusual for vehicle design, providing asymmetry in the shape.

Not all consultancies are the same, and each has its own personality and way of doing things. The Heatherwick Studio emphasizes a strong commitment to collaborative enquiry in a spirit of "curiosity and experimentation" and has a particular emphasis on large-scale and unusual projects. Heatherwick himself denies the studio has a fixed style: "it is more like solving a crime. The answer is there, and your job is to find it. So we go off and do bits of research that essentially eliminate suspects from the enquiry. And you follow up leads and gradually narrow down the potential solutions. Ultimately what you're left with is the answer."

Many companies might perceive a need for a new product and seek to improve an existing design. Product designers are trained to "know" that there are new ideas to be had and have the drive needed to research and the confidence to question, challenge, and ultimately break new ground. This is the real strength behind many of the top designers and design consultancies in the world.

1.24

Figure 1.24
Olympic Cauldron. The realization of this innovative idea for a shared, flower-like lighting of the Olympic flame involved hand-beating copper sheet over MDF formers to produce 204 unique petals in a design that was around 20 times lighter than the cauldron used in the Beijing Olympic games.

Chapter summary

This chapter has highlighted the role of research as a requirement before setting out to design a product. It has done this in three distinct but related sections.

It began with a section exploring the role, thoughts and behaviors of people in relation to a specific project or problem. If key user insights can be recognized, then there is already scope for generating a creative new product. The next section continued by adding in the general or specific requirements that any developed product must have in order to be safe and compliant with regulatory requirements. The chapter concluded by emphasizing the cognitive skills needed to appraise and communicate this information in a product brief.

Some designers argue that design is a more individualistic approach than the processional process described here, with ideas perhaps developed through the intrinsic processes that revolve around individuality, independence of thought, experiment, reflection, and empathy. Certainly there are arts and crafts designers who can do this on a small scale, and even some designers who can command large sales through invention. To be a professional product or industrial designer requires consistency and an ability to generate good products time and time again. Guaranteeing this level of success on the scale needed for mass-produced products, however, requires this form of process.

It is also sometimes argued that being led by user needs leads to "mundane" outcomes because ordinary folk simply do not know what can be achieved—sometimes referred to as market myopia. The secret is to be guided by the research, and not ruled by it.

Assignments

1. Imagine you have been asked to curate an online exhibition of the most influential products in the past ten years. Collect and display your products via a series of images. The exhibition theme should be products that have helped to enhance the quality of life or enacted social change.

2. Video a group of people undertaking a simple task such as making a cup of tea or cleaning their teeth. Even with friends or family, you should still seek their approval for this. Then watch the video many times over. Your aim is to observe some of the interesting rituals and insights taking place. Recording whether the majority of your insights come at the beginning of your viewings or at the end. Try then to design a product that takes advantage of your observations, perhaps a new kettle design, cup design, or toothbrush.

3. Mobile phones have moved through distinct changes: large analogue, small analogue, large digital, small digital, hinged digital, twin hinge, folding, and sliding. Use scenario analysis to consider what the design of a mobile phone might be in ten years' time? Similarly, what might be the analogous development of a portable vacuum cleaner device?

4. Some of the findings attributed to cool hunting techniques include the statement that the more developed a culture becomes, the more eccentric it gets, or that everyone wants to keep his or her options open. Consider one of these two statements and consider how you might redesign a domestic kitchen wastebasket with this in mind.

5. Investigate a range of new and emerging technologies by studying academic papers or science journals. Propose a range of problems that the technology might solve complete with draft products.

6. Produce a set of twelve criteria by which you might benchmark mobile phones. Your criteria might include the ease of pressing the buttons, the ease of using the menu, and the ability to adjust the volume. Conduct a series of tests on a range of mobile phones to consider which one might be the ultimate performer. After your benchmarking analysis, reappraise your criteria and order them by priority according to their importance. Repeat this for other ranges of products such as writing pens or bicycle padlocks.

Product concepts

This chapter aims to show how designers turn the product brief developed in Chapter 1 into an idea or product concept. Being able to generate ideas is what many people perceive product design as being all about, and it is also what many young designers claim they are good at. It is, however, surprising how often mundane thinking emerges under the pressure of a design project. The chapter starts by looking at some of the mechanisms behind creative practice, giving insights as to how it works and how it can be used at times to solve complex design problems. The second section shows that ideas alone are not enough and highlights how creative ideas still need to be worked through to generate a truly winning product concept. The final section illustrates some of the market trends and outputs of creative practice from current designers developing new and innovative products.

Generating ideas

Creativity is one of humanity's greatest capabilities. Despite its importance in our evolution however, it's surprisingly hard to agree on a common definition of creativity or to understand exactly how it works, and it can at times be frustratingly elusive. An ability to generate new ideas time and time again for any number of different projects is, a defining feature of a "professional" designer as opposed to a casual inventor.

To be more creative, or creative at will, it's helpful to move away from models that see creativity as being a point or moment of inspiration, to models that see creativity as approaches to problem solving involving a number of steps such as preparation, incubation, illumination, or validation. This first section on creativity lists each of these areas and describes some of the behaviors and practices that can then operate allowing creativity to flow at every possible point.

Creative preparation

Being open minded

It can be difficult to take a different view if the life you lead is closeted or routine. Imagine growing up and living within an enclosed room where everything inside was blue. If you were asked to design something new, your product would inevitably be blue. If you keep the same customs and habits, hold the same thoughts, and see the same sights then you are living in your own blue room, and your capacity to learn about, compare, and contrast ideas becomes limited. In fact, set patterns of thinking can actually create rigid neurological patterns that physically restrict our mental abilities to be creative.

Look with an observant eye at life around you, not just at the products but at the everyday details. Some designers use sketchbooks or photography to capture ideas and thoughts in simple forms while they are working or on the move. Being open-minded involves the difficult concept of accepting that everything you know or have learned might actually be

"Imagination is more important than knowledge."

Albert Einstein

2.1

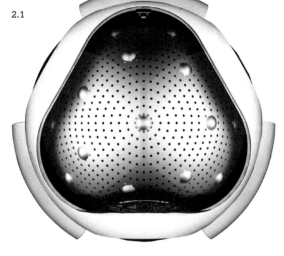

Figure 2.1
Maglev concept
washing machine
by Jakub Lekes

wrong, and that everything other people do, say or know might also be wrong. Being open-minded also involves overcoming the belief that every problem must have one, optimal, best-fit solution to a problem. This belief of one best solution may be a generic human trait, but it is one that is almost certainly exacerbated by our logical and analytical development through schooling. Designers again know this is not the way to approach a problem and that there may be hundreds of different solutions. One simple way to implement this approach is to explore as many workable ideas as possible rather than settling for a few.

Related to open-mindedness is the notion of *serendipity*, a term that describes discovering things by chance. It is said that as many as 18 people may have had the opportunity to "discover" antibiotics, but it was the recognition that mold spores might actually be useful that lays the discovery of at the door of Alexander Fleming. Another example is Percy Spencer who is reputed to have noticed a gooey mass of chocolate in his pocket while working on ways to mass-produce Magnetrons for wartime radar systems. Many people might simply have been annoyed with losing their chocolate and ignored this seemingly useless phenomenon, but Spencer realized that the magnetron must have melted the bar so he repeated the experiment with corn and eggs. His employer, Raytheon, ultimately proceeded to develop the world's first microwave cooker.

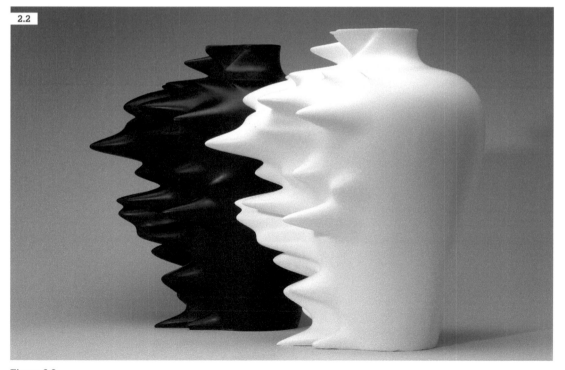

Figure 2.2
Hyper Fast Vase by Cedric Ragot. A solid surface molding technique is used to express the idea of a Ming vase frozen in a state of acceleration.

Exploring

Exploration is a proactive way of being open-minded. This might just mean asking What if . . . or How can . . . questions or simply talking to other people. It might mean playing with the Internet, which can make a whole world of knowledge accessible within seconds. Exploration might also entail new experiences such as traveling or taking up new hobbies or simply reading a different newspaper, talking to someone new, or sitting in a different seat.

Perhaps best of all though, it might mean physical exploration: finding and working with materials, shapes, and patterns; taking things apart to see how they work; or building mock-ups and models. These do not have to be precise or purposeful but simply a mechanism to help you build up a library of new thoughts and knowledge. Try also doing things differently. Try pulling a wheelbarrow instead of pushing it, lacing your shoes in a new way, reading a newspaper backwards, eating breakfast at night and dinner in the morning, or using tea leaves instead of tea bags. Perspectives that come from these activities can spark new insights, new trains of thought, and new ideas.

"Hell, there are no rules here, we are trying to accomplish something."

Thomas Edison

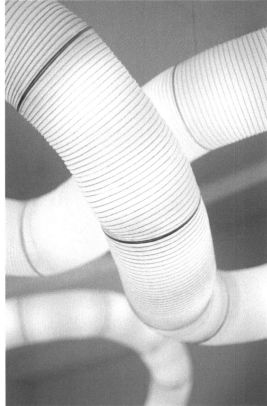

Figure 2.3
Teiko by Anthony Dickens. A prototype modular lighting system inspired by Japanese paper lanterns; it can be adapted to transform spaces.

Playing

As people get older, their abilities for abstraction and analysis increase allowing learning to be more didactic. Because children find this instructional approach challenging, tiring, and boring, they learn more through a variety of more practical routes. Designers should have no qualms in reverting to a play mode in much the same way that young children learn. It can be more illuminating to experiment, explore, mimic, or make something with no purpose than to think in pure abstract or to try to apply knowledge according to rules and logic.

An adult, for example, might never make the connection between a car and a washing machine to make a washing machine car, but a small child has no problem in doing this or any other number of seemingly illogical thoughts. But why not have a small washing machine in the car? There is water, heat, and motion for tumbling and drying. Sometimes the ideas go nowhere, but occasionally a new breakthrough will appear.

Having fun through play also reduces stress, and being more relaxed can help the thinking process. Design-centric organizations recognize this too and may have spaces for staff to play or encourage playful activities.

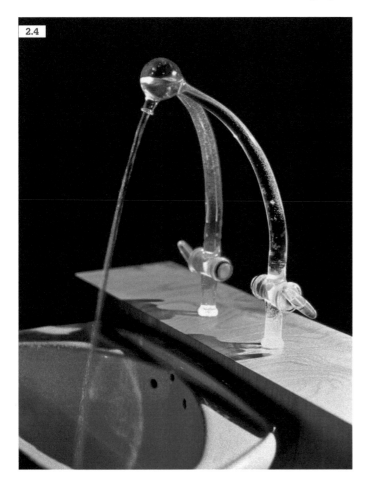

2.4

Figure 2.4
Glass Tap by Arnout Visser for Droog.
Much of Visser's work draws on his reflections upon physical or mechanical laws and on traditions of form and function. The resulting product often incorporates subtle humor or design twists, as is evident in Visser's Glass Tap.
Photograph: Erik Jan Kwakkel

"I have tried to develop my own approach to design and my own idea of what design is or, more interesting, what it could be. My main interest is in making new and different stuff by using the knowledge I collect as a hobby and mainly about technology and how we relate to it."

Mathias Bengtsson

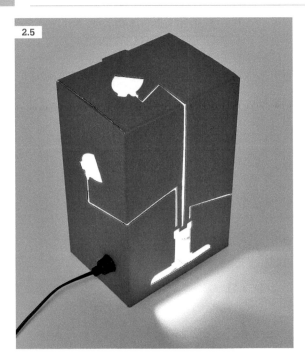

2.5

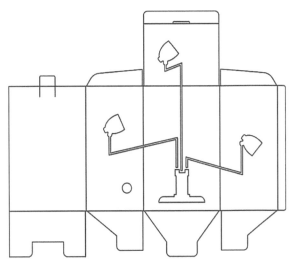

Reflection

Most people will recollect an inspired thought that has often come during moments of tranquility and relaxation, perhaps in the bath, in bed, or at a bus stop when there is nothing else to think about. Our modern lifestyles mean there are immense time pressures to act quickly, eat quickly, pack the day with activities, and find an idea quickly to keep clients and bosses happy. This "busyness" can keep us from having the time to experiment and to learn or play, cutting our opportunities for imaginative thought. It can also stop the mental process of thinking about and reflecting on life all around us.

Often, therefore, reflection means just taking some time to rest in order to be able to think, freewheel, daydream, reflect, and let your imagination work. There are, however, many activities that will stimulate reflective practice, and these might include practicing meditation or yoga or simply taking a regular walk. Even sleeping can refresh the brain's synapses, enabling new thoughts to take place. Even though this might seem relatively straightforward, it can be counterintuitive and often takes determination to actually stop and take a reflective break amid the frenzy of life.

"In design sometimes one plus one equals three."

Josef Albers

Figure 2.5
Not a lamp by David Graas. Not a lamp incorporates a silhouette of a light in the lamp's shade. It's this silhouette that glows rather than the space around it, offering a subtle but unusual perspective. The shade is assembled in a parody of flatpack furniture (as shown in Figure 2.14b).

Incubating ideas

Adaptation

It is sometimes thought that all ideas for new products must be radical, but in reality most new products are just subtle improvements on previous ideas. Nevertheless, quite profound design effects can be had by simple adaptations to existing designs such as changing scales, altering features, or redefining attributes. The creative tool SCAMPER (*S*ubstitute, *C*ombine, *A*dapt, *M*odify, *P*urpose, *E*liminate, *R*everse) is often used to explore design adaptations.

Analogy

Drawing an analogy is making a comparison between things, and is a key feature behind many approaches to creativity. What might the solution to your design problem be if we contrasted it to the solutions derived in another, similar context? For example, an airplane is like a bird in that they both fly, have wings, can travel for a long way without landing, and can sense where they are going. They are unalike in that they fly through different means

2.6

Figure 2.6
Mezzadro Stool by Castiglioni. The Mezzadro Stool is an aluminum tractor seat attached to a chrome-plated steel stem with a beech footrest. It is an example of a ready-made drawing of everyday objects to make design pieces. The adaptation of a tractor seat has been lauded for its expressive strength, starting as an exhibition piece before being taken up for production by Zanotta.

2.7

Figure 2.7
Opto reading glasses by Xindao B.V. These magnifying glasses are analogous to reading glasses in the way they work, and the rounded bottom is analogous to an egg allowing it to stand in an easy to grab way.

2.8

Figure 2.8
Graphene aerogel. Aerogel is a low-density solid with fractal molecular structures making it strong but almost as light as air. One form is a hazy blue sponge-like substance that is made from the same material as glass but is formed from 99.8 percent air making it 1,000 times less dense. Graphene aerogel shown in the image is amazingly light material with 1 cubic meter weighing about the same as an apple. Carbon-based aerogels also have high surface areas and high absorption of heat radiation. What problems might you resolve using these materials?

and work in different ways. Therefore, if your specification is related to the development of airplane wings, is there a solution that can be derived from birds? If your problem was how to innovate the design of an iron, you might consider how making things smoother is done in other contexts, such as making pastry flat or making freshly made roads smoother. Analogies can be natural, personal, remote, or fantastical and are often just used very informally: "This problem makes me think of X (analogy)—that suggests to me that maybe we could try Y (idea drawn from analogy X)."

New materials

Designers sometimes consider materials late in the design process, simply selecting the most appropriate material for the application. In fact, just to stimulate ideas at the front end of the design process, a designer can explore thousands of different materials with an incredible range of characteristics and performances. Many companies house a library or box of materials for just this purpose. Exploration might be through a tactile or visual appearance. For example, brushed aluminum has a smooth, satisfying grain but is easily marked with greasy fingers. This can be annoying for, say, the design or a kettle—or is it? What other product ideas might this inspire?

The vast range of emerging materials makes exploration even more relevant. Developments in nanotechnology, composites, biocomposites, gels, foams and other treatments offer designers incredible new opportunities.

New technologies

Technology can be an elusive term to define, but it might broadly encompass the knowledge and application of tools, crafts, and science that we use to shape the world. Hence, although it would include common associations such as information technology or the advanced high technologies of the aerospace industries, it might equally include notions of systems, methods, and, in fact, anything else that brings about change. Like new materials, therefore, new technologies have the capacity to inspire product ideas on a massive scale.

The issue for designers is often being able to keep abreast of or access new technological developments. Many will be confidential or lie hidden within company research labs with potential that remains unrealized. Networking events, trade exhibitions, science magazines, technology transfer and patent sites, and partnerships between designers and technologists might release new opportunities by helping to outline and explain technologies.

2.9

Figure 2.9
Segway by Dean Kamen. Despite having just two wheels, Segway transporters are able to self-balance and remain upright, providing a novel way of steering and traveling. This is achieved through the development of "dynamic stabilization" technology revolving around five micro-machined gyroscopic sensors. These solid-state (silicon) devices use the Coriolis effect to determine the angular turning rate, which provides feedback information to the motors allowing corrections to be made and keeping the device upright.
Photographs: segway.com

Illumination—Breakthrough ideas

Subversion

If it can seem hard to generate new ideas within the normal rules, then break the rules. This might mean breaking social norms of etiquette or behavior or challenging rules, including the rules of nature or physics. This can feel like cheating, or it can feel utterly subversive, but it can lead to far-reaching new thinking and design. The green gunk that cleans dirty workshop hands, for example, started life as failed protective barrier for products such as metal and silk stockings.

Figure 2.10
Massive Infection by Pieke Bergmans. Young Dutch designer Pieke Bergmans's favorite design methods include subverting existing production processes to create new forms and functions, which she then describes in terms of virus or infection. Bergmans's claims that her designs stem from an unstoppable curiosity.

Scenario analysis

A commonly applied approach to generating a creative design solution to a brief is to consider how the design problem you are working on might look in different scenarios. This can range from the simplistic, such as how it might work with different gender or age groups, to more challenging scenarios, such as how might it work or look in the future. What will people want from your product in ten years' time, which you might work toward delivering now? Are there any links or common "forces" or issues across the range of scenarios? Are there any seed trends (i.e., what are the current trends that might develop in future)? Are there any parallel scenarios or precursor trends (i.e., trends based on similar or previous behaviors)? Thinking through several scenarios with a panel of people with mixed backgrounds is ideal. The intention is to squeeze out ideas, insights, or trends that might translate backward into the current design.

2.11

Figure 2.11 Morph device concepts by Nokia. The Nokia Morph is a concept that demonstrates the functionality that nanotechnology might be capable of delivering, including, for example, flexible materials, transparent electronics, self-cleaning surfaces, and the ability to harvest its own energy.

"I think that good designers must always be avant-gardists, always one step ahead of the times."

Dieter Rams

Lessons from nature

The flora, fauna, and patterns of the world around us have been evolving over millions of years so there are great lessons and ideas that can be learned from nature. How do owls fly silently? What are fractals? Why don't fish in swarms bump into each other? The term *bionics* is used to describe the application of nature's biological and botanical methods and systems, such as flocking, nesting, hunting, swarming, preening, and defending, to the study of engineering and design.

There are many examples of bionics in action: rippled skins rather than smooth boat hulls replicated from shark skin enabling more efficient transmission through water; echo location of bats mimicked in sonar or medical scanning; and the strength of the small plant burrs that stick to animals and clothes, which has inspired the creation of Velcro.

"Yves Béhar is a thinker. And one thing he thinks about a lot is the future."

Herman Miller

**Figure 2.12
Grey Heron hair
dryer by Eliran
Yaksein.** A hair
dryer with unex-
pected surfaces
and relocated
features inspired
by the heron bird.

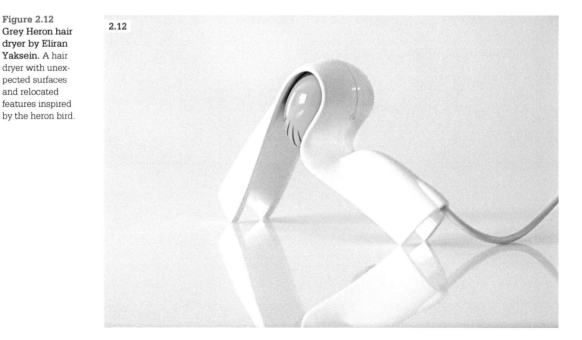

2.12

New links

Creativity often involves making links between areas that have no normal associations. When designers repeat these links across a range of products, they can develop distinctive career styles. For example, Raymond Loewy, an elder father of classic industrial design for example, made much by linking the (new at the time) technology of streamlining to the style of products. Dieter Rams's attempts to link timelessness with products, or "everlasting design," has produced trademark sleek and minimalist products. These styles can even become iconic genres when they become de-rigeur.

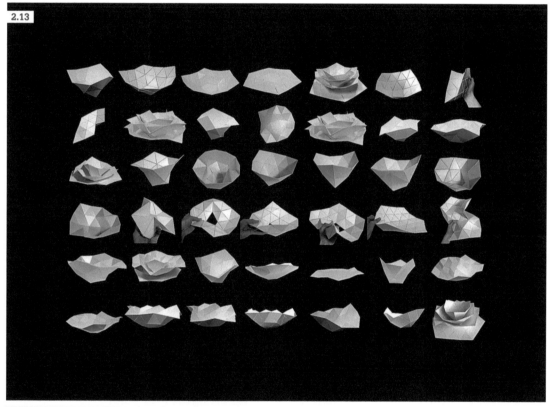

2.13

Figure 2.13
Crushed bowls by Julien de Smedt/Maria Ljungsten for Muuto. The bowls retail through Danish company Muuto and have applied the same principles of equilateral triangles used in the computer-aided design of architectural projects to the design. The bowls are made of bone china and are unglazed and hand-polished on the outside and glazed on the inside.

"Ah good taste! What a dreadful thing! Taste is the enemy of creativeness."

Pablo Picasso

Validation—Being bold

Taking a different view

If people simply copied the work of others, then nothing new would ever emerge. Even just thinking in the same way would result in products that were broadly similar. Some of the world's best innovations have come not from experts but from ordinary people who are unconstrained by convention. For example, Alexander Graham Bell is credited with inventing the telephone but was not an electrical engineer. The great inventor Thomas Edison is quoted as saying that "if Bell had known more about electricity he would never have invented the telephone."

Figure 2.14
Bubble storage unit by Aziz Sariyer.
Sariyer's aluminum bubble storage unit is lacquered on the outside and painted on the inside with phosphoric primary colors. It has four clear glass shelves, making a practical product; however, because it is nearly 2 meters tall it offers a different perspective by providing shelving that also has a standalone, sculptural quality.

2.14

Thinking against convention, however, does not necessarily come easily. There is a natural herd instinct to follow the crowd among most people. It's safer, and there may be a presumption that everyone else might be right or that if things have always been done that way, then that must be the best way of doing it. Designers rarely accept the idea that things either can't or shouldn't be changed. This does not mean being contrary or argumentative, but it does mean being prepared to question everything, being able to hold a different view,

and being able to voice an alternative opinion. The challenge often involves referring back to the product research and the brief, but not being constrained by it, to work principally with the main problem rather than trying to solve the whole brief in one hit, and to not be constrained by a fear of what people will think.

This can consequently produce radical new ideas, or simply obvious but unidentified ideas that still seem fresh and exciting.

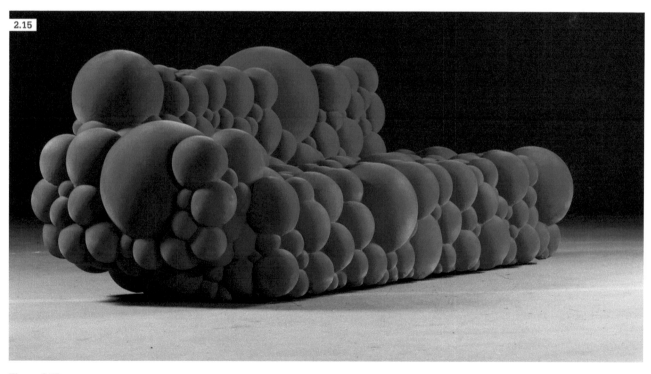

2.15

Figure 2.15
Modern organic furniture by Maarten de Ceulaer. Part of the mutation series, the sofa design explores the natural concepts of cloning and growth within a furniture context.

Seizing opportunities

Generating ideas is one thing, but being able to recognize the validity of a new idea, and take the risk in it is something completely different. It takes a special type of mind and courage to seize new ideas. The penalties for failure can be severe including the daunting risk of ridicule and financial failure. Great designers are rarely timid; instead, they are ready to tackle new challenges, work with new people, and try new things.

Creative techniques

It is possible to be creative without having any of these inherent creative behaviors because there are creative tools that can do the same job—effectively "forcing" our brains into more creative thinking. The Gordon method, Occam's razor, brainstorming, and TRIZ are just some of the many hundreds of these creative tools.

EXAMPLE TECHNIQUES

The Gordon method is a creative technique in which the participants try to solve the principle of a problem without knowing the exact background or context to the problem. Occam's razor similarly "shaves" away unnecessary information. Both techniques can help to avoid the biases that can come from less important, mind-cluttering problem context and detail.

Probably the best-known creative tool is brainstorming, a technique involving people generating ideas in rapid, open, and free group sessions. It is probably popular because it is simple to do, but in reality it is a technique that is often done poorly where insufficient time is allotted to the process or awkward group or power dynamics are involved.

TRIZ, or the Theory of Inventive Problem Solving, is a more recent tool. Based on the observations of the patenting system by Genrich Altshuller, it asserts that there are distinct patterns of solutions for different problems. Therefore, it's possible to have an algorithm that generates a generic range of solutions for any particular problem.

Six hats, snowballing, morphological analysis, mind mapping, brainwriting 6-3-5, SCAMPER, and IDEO method cards are some of the other commonly used tools that are worth getting to know.

2.16

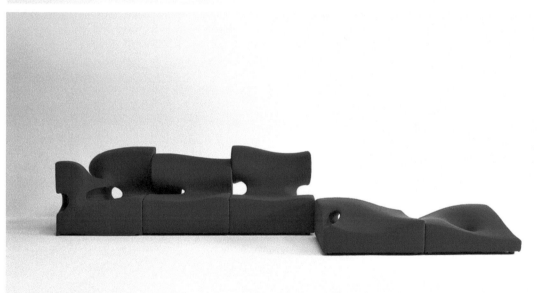

**Figure 2.16
Moroso Misfits
by Ron Arad.** The idea of putting together a modular seating system with ill-matching shapes might not be a consideration for most designers and, if it were, it might challenge natural sensibilities and be quickly and easily rejected. Yet Ron Arad's working of lines, curves, and spaces produces a seat with a satisfying array of undulating movements.

Case study
Wayne Hemingway MBE

Wayne Hemingway achieved initial success with his partner Geraldine through their individual fashion designs, which were sold from a stall in London's Camden Market—an enterprise that ultimately led to the launch of the Red or Dead fashion label. The label was lauded for its streetwise style and its willingness to challenge both the established fashion industry and the media on frequent occasions. Selling the business for multiple millions might have been a cue for retirement, but Hemingway subsequently engaged with a number of architectural and housing projects, challenging established thinking and winning numerous awards and design contracts in the process. His canon of work includes products such as digital radios and ceramic tiles, and his clients include Sky Plus (Sky+), Wanadoo, British Ceramic Tile (BCT), Boddingtons, Sony, Royal Mail, and the Caravan Club.

Hemingway does not have a formal education in product design, and his ability to innovate across a variety of fields is a testament to the way in which a designer can identify and offer solutions for contemporary challenges through passion, a willingness to explore and reflect, an eye for detail, and a drive to seize opportunities. The Shackup, HD Roadrunner, and Butt Butt projects have to be seen in the context of the designer's interest in the way in which people interact, occupy, and work with space and with each other.

The Shackup is Hemingway's response to his observation that sheds have moved on from their traditional storage and potting functions and now serve as spare rooms, work spaces, reading areas, or studios, all of which are much more lifestyle-orientated purposes.

The Shackup range includes a small shed designed purely for bike storage and a range of sheds that combine traditional storage facilities with outdoor office space. This flexibility is enabled through a high roof height, large glazed areas and solid construction. Hemingway provides the bike too, and his Roadrunner folding bike has been one of the cheapest folding bikes on the market.

Figure 2.17
HD Roadrunner. This bike is designed to be the cheapest fold-up bike on the market and to provide an opportunity for householders to be able to afford, store, and use a sustainable method of environmentally friendly transport.

He even helps to create the cycle paths as a patron of the Sustrans cycling movement.

Hemingway's more recent work has explored both events and products that draw on and instill a sense of vintage and popular culture. Combined, Hemingway's ideas on architecture, space, living, and travel represent a man striving to deliver his vision of the world through the medium of design.

An interview with Wayne Hemingway

A good designer seems able (if necessary) to cut across all of the practices, traditions, and clutter of life to get to the heart of an issue. What's your secret?

Being able to identify problems, things that can be improved, gaps in the market . . . and, importantly, being dissatisfied with things and having a questioning mind—one that rarely shuts off.

You seem to have experimented with style and culture in your life. Do you think this helps?

I started going out to clubs when I was thirteen and grew up in the golden age of youth culture when we all experimented with soul, punk, rockabilly, ska, mod, new romantics. We dressed in the clothes, watched the films, researched the music . . . it had to help.

You have been able to find new ideas in the fashion, building, product, and archiving industries. Do you find you need to adapt and change your ways of working?

I've been very lucky in that I have had opportunity and freedom to go where I please. The basic ways of working don't change across the industry . . . identify need, understand customers, find a great team to help deliver, work bloody hard and market your concept wonderfully.

With Red or Dead, and the Land of Lost Content, you've forged your own path. Do you find working for clients such as B&Q, PURE, and George Wimpey more restricting?

More challenging, yes. It's harder in that there is a lot of cajoling, persuading, explaining and compromising to do, but we learn lots. It's an opportunity to reach a large new market and economies of scale help get our ideas to a wider audience. That's very important to us, it makes the brain work!

Do you ever look toward new materials and new technologies for inspiration?

That's an integral part of being a designer. Having a knowledge of what's new and bringing that into your work.

2.18

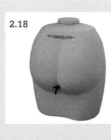

Figure 2.18
Butt Butt for Straight LTD. A water barrel, known in the United Kingdom as a butt, here modeled as a butt! Aside from this pun, the butt has a flat front allowing it to be placed flat against a wall. It has a clear separation from traditional water barrels that will win it a unique target market.

You have taken a few risks in your career; do you ever worry that you are going to fail?

There's always a chance of failure, but we have become good at balancing risk, and fortune favors the brave. As long as you set out on a project for the right reasons and put as much as possible into it as you can, then failure is nothing to be ashamed of.

2.19

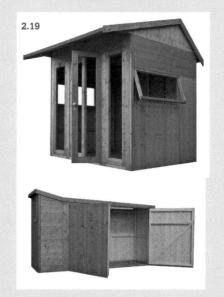

Figure 2.19
The Shackup bike store. Hemingway's store is made from sustainable timber and designed to lend itself to bike storage, a concept that seems obvious but is rarely seen in practice. A key concept behind the Shackup shed range is multifunctional purpose. It is tall and solid and has large doors and window areas to maximize the flexibility of the space. It might feasibly be considered an external house extension rather than a traditional garden store. The shed shackle is a further element to the range that provides an additional level of security for shed contents, again recognizing a social trend—this time, the increasing level of theft from garden areas.

Nurturing ideas

An ability to get all of the ideas generated down into some pictorial representation and committed to a permanent recordable form is now fundamental. It is not often that an idea will be perfect from the outset so ideas will need to be worked up so that each one is developed and improved until it is capable of being evaluated in its best and most viable form.

Sketches

Some designers at this ideation stage will reach for their CAD or graphics package particularly if they are not confident at drawing, but most tutors will probably tell you this is a mistake. Sketching allows you more freedom to clarify and test thoughts, provide new perspectives, and play with and explore ideas in an individualistic way that CAD does not. CAD can also take time and investment and lead to impressive visuals that preoccupy designers on the image rather than the quality of the idea. The process of externalizing ideas through sketching means you don't actually have to be good at art so most design sketches may be simple pencil line drawings. The clearer you can make your sketches, however, the more clarity you can provide, and the better finish and **render** you have, the more credibility you may get in discussing your ideas with others.

Levels of ability in the skills of observation, interpretation and hand eye co-ordination can be practiced and improved with more time on a sketchpad (and can be done even on the move if you carry this with you). It's also relatively easy to improve the appearance of sketches using simple techniques that can create a more professional 'design style' appearance. Stronger line strengths to highlight outlines, simple shading to lift a sketch of the page and provide three-dimensional depth, marker pens to embellish color and lighting.

Figure 2.20
ZernO by man-
worksdesign. Sit?
Lounge? Recline?
Talk? Watch? The
ZernO explores
and offers ideas
for the way that
people and sofas
interact.

2.20

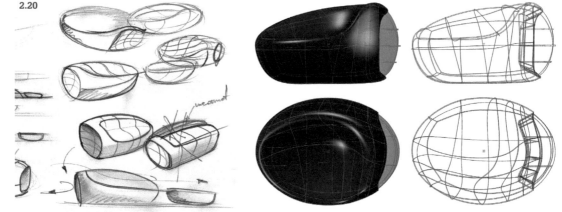

"A picture is worth a thousand words."
Proverb

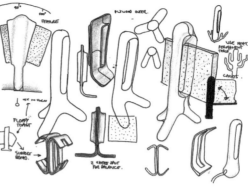

Figure 2.21
Glide ceramic toaster by George Watson. There has been little development of the toaster since the start of the twentieth century. Other appliances have been developed and improved, incorporating new technologies and thinking, but the toaster has remained relatively untouched. When it was first invented, eating toast was a social activity that took place at the breakfast table. Today, toasters have been relegated to low-cost artifacts hidden away in the kitchen. George Watson's Glide ceramic toaster is designed to engage the user and reinvigorate the social context of breakfast.

PROJECTIONS

Projections are different ways to present a 3D object in two dimensions. Lines that appear parallel but that actually converge onto a distant point provide depth to a drawing. This form of perspective drawing is referred to as an iconic projection and adds a greater sense of reality to drawings.

Oblique or isometric projections are referred to as axonometric and are rotated along one or more axes presenting an angled view.

An orthographic projection displays a three-dimensional product in a two-dimensional way through its top (plan), front and side views. It is used in manufacturing but can have a role to play in helping to map out the basic features behind a product at the concept stage.

"Designs of a purely arbitrary nature cannot be expected to last long."

Kenzo Tange

Figure 2.22
VAX development sketches

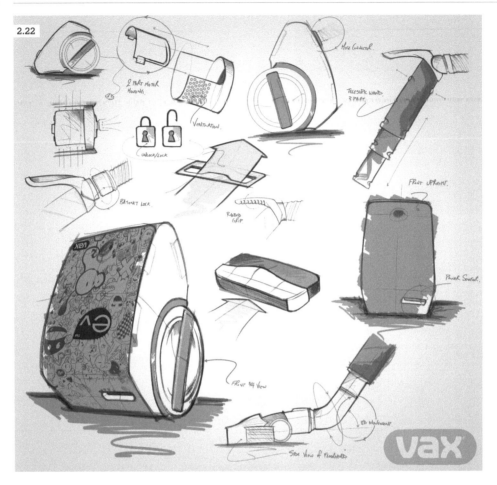

Figure 2.23
VAX cardboard cleaner. The VAX cleaner concept is made from recycled and recyclable materials and has been designed by Jake Tyler in collaboration with Vax.

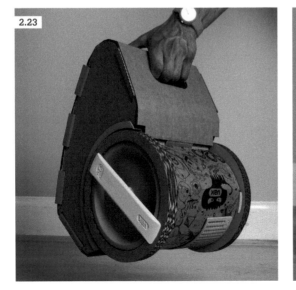

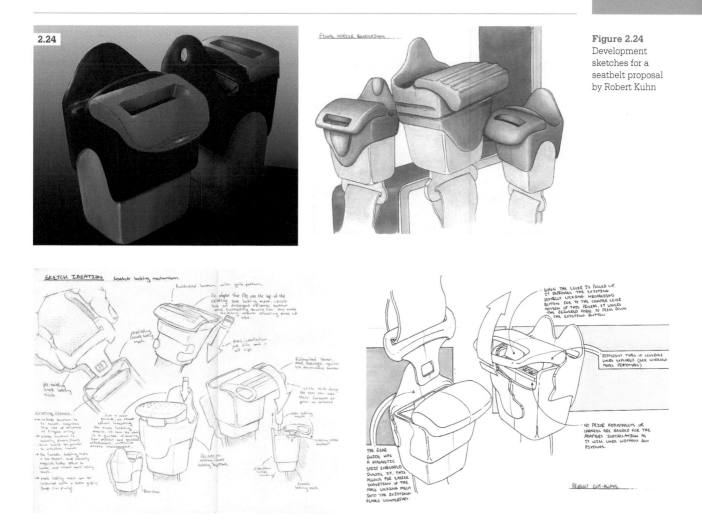

Figure 2.24
Development
sketches for a
seatbelt proposal
by Robert Kuhn

"We meet the [designs that accommodate all their functional
requirements] by evaluating a series of alternative designs.
When I begin a new project I will usually, but not always,
start with hand drawn sketches, using colour media . . . through
the balance of the project I gradually add more details."

David Sharbaugh, Point by Point

Concept models

The creation of physical models is another vital tool in getting the true feel of a concept. Models can be made quickly from inexpensive from card, foam, or clay. The building and subsequent exploration of the model can enable profound understanding of the way a concept might evolve in a way that even sketches can't. It can, however, be easy to get carried away with modeling, in the desire to produce something that looks fantastic at regular stages of the design process. These can be expensive, time-consuming, and totally inappropriate.

Hence, it is very important to be able to select the most appropriate and cost-effective model to suit the stage of the design. First and foremost, a model is created to help predict the future of an idea. It can help to shape the

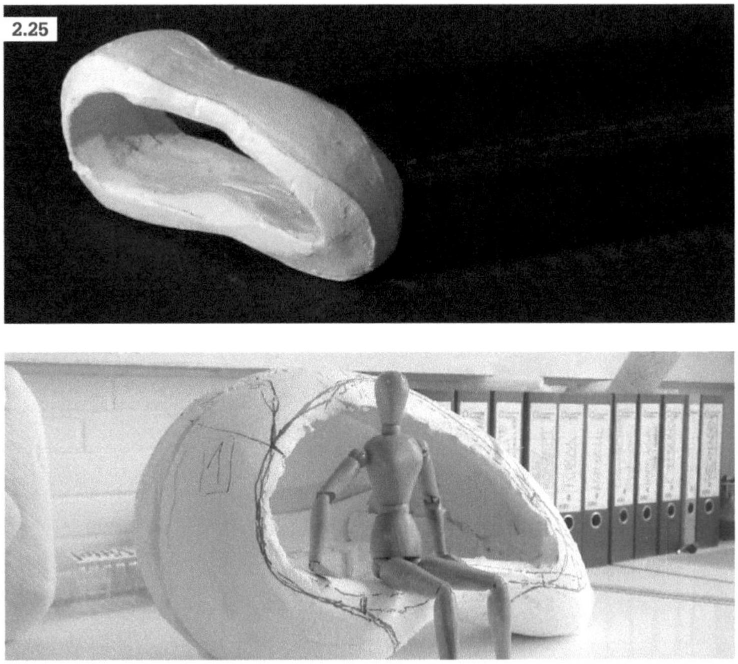

Figure 2.25
Foam, scale development models used for developing the idea of the Napshell power-napping bed

"I tried a dozen different modifications that were rejected. But they all served as a path to the final design."
Mikhail Kalashnikov

thinking behind the appearance, performance, and expectations of a design. In addition to shape, models cast in resin or made from clay or extruded polystyrene (blue foam) have been the traditional and most effective way of testing ergonomic feel. At this conceptual stage, a well-made model constructed of clean white card can achieve greater impact than one that attempts unsuccessfully to match surface texture and coloring. Old toys, modeling and toy construction kits, and broken equipment are ideal for simulating moving parts. Having a box of bits to play with can build up engineering know-how and provide inspiration for future projects. Mathematical and other symbolic models can equally help to forge an understanding of how a design might work.

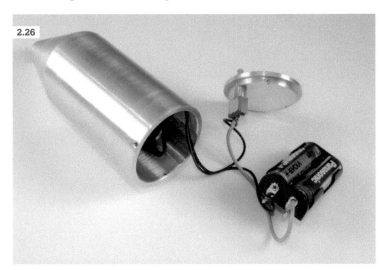

2.26

Figure 2.26
Blaze bike light by Emily Brooke. The Blaze bike light is a good example of open-minded problem solving. It uses a laser to project an image of a bike ahead of a cyclist to warn other road users of the cyclist's presence rather than simplify making the cyclist brighter through more lights or reflective clothing. This helps to reduce the number of accidents from vehicles turning across cyclists, the main cause of cyclist fatalities. Iterations of the laser light from this early stage model have turned a clever concept into a multi-award-winning design.

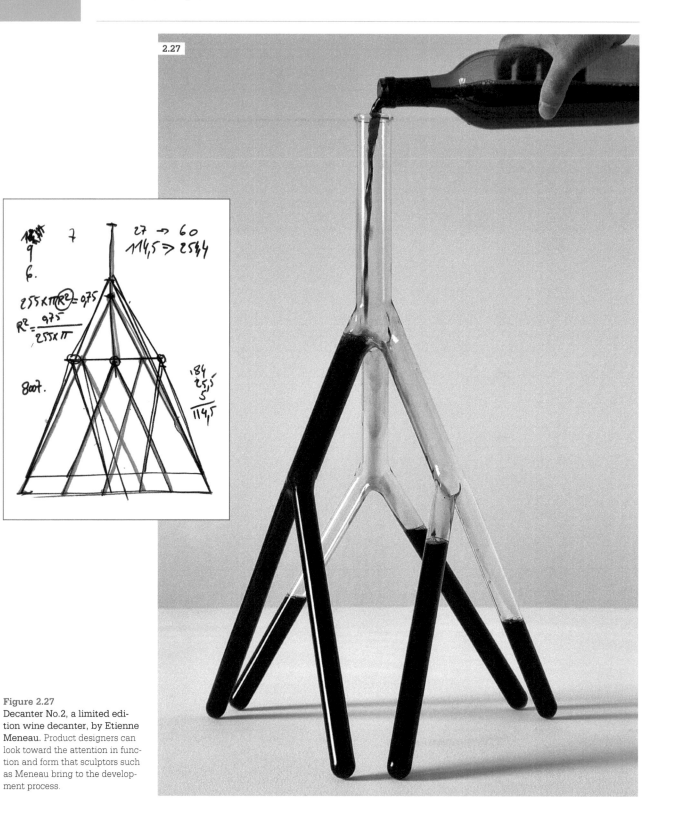

Range of Ideas

The exploration of ideas through sketching, modeling, reflection, and evolution should aim to develop a wide range of possible solutions to the design brief. The emphasis is on range; one idea presented in four different colors is not a range, nor are two or three ideas that work in the same way but are simply packaged differently. The range of ideas should present alternative methods to resolve the problems, low- and high-cost alternatives, simple and complex conceptions, obvious and nonobvious plans, and possible and less possible suggestions.

It's also important to play with each different idea. The time and effort taken to evolve an idea into a more developed form demands time and mental strength because further sketching, discussion, and reflection can seem unnecessary when you already seem to have an idea that works and where there are business pressures to find solutions quickly. However, there is nearly always scope for developing an idea through to a better state.

2.28

Figure 2.28
Lenee microwave washing machine. A concept design for Electrolux by Andrej Suchov. Microwave technology might not just clean clothes, but it could be used to repair or transform clothing colors and textures.

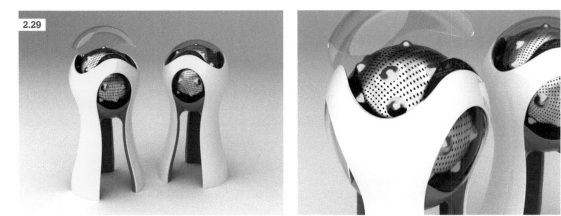

Figure 2.29
Maglev washing machine. A concept design by Czech designer Jakub Lekes that overcomes the problems of vibration, noise, and bearing wear in conventional washing machines by using magnetic levitation to suspend a rotating drum. In theory, the spherical drum could rotate faster and in more directions than a conventional rotating drum allowing for more efficient and effective clothes cleaning.

Innovation

A range of ideas should always seek to be new and exciting, not just meeting but exceeding market expectations. This might include both small scale (incremental) and large-scale (radical) innovation. Large-scale innovation representing quantum leaps in thinking and design are undoubtedly exciting, and there are many examples of inventions that have changed and sometimes shaken the world. Organizations that release a radical new product are shown to maintain a significant market share even after competitors catch up, but there are also other long-term benefits including spin-off products, a culture of more success, and a goodwill factor with the public for being seen as innovative. Normally, radical innovation is the result of a design team hard at work within a commercial organization. Individuals within that team, associated consulting organizations, or designers are all capable of reaping the kudos that radical and successful new products bring. This is often the dream of both new and experienced designers. It is, however, often not possible to generate revolutionary new ideas every time, and as few as 5 percent of new products released into the market might actually be considered radically new.

2.30

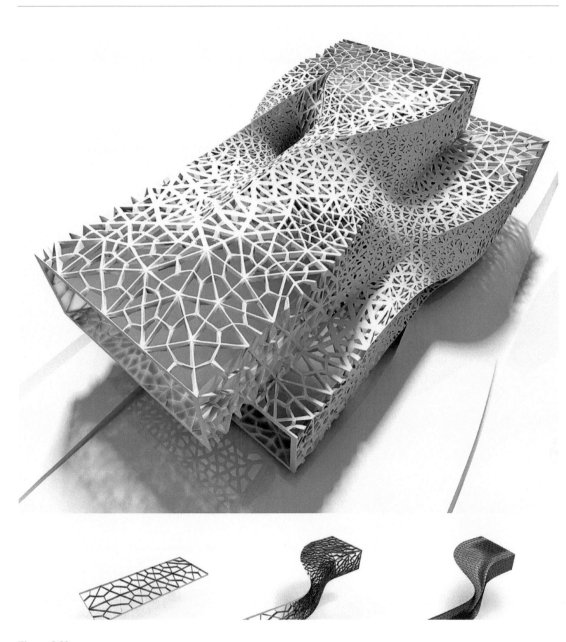

Figure 2.30
Jellyfish House by Iwamoto Scott. The Jellyfish House includes a range of sensors that enables the skin of the house to monitor light, heat, and water—even in a contaminated environment—and to react accordingly. The project draws on digital modeling and finite element analysis of a phase-change material to achieve a structure that fluctuates between solid and liquid states.

Case study
Naoto Fukasawa

"My products are already in your mind, you just have not seen them yet."

Naoto Fukasawa

Naoto Fukasawa was born in the Japanese prefecture of Yamanashi in 1956 and studied at the Tama Art University where he now teaches alongside running his own design consultancy. He has a truly outstanding talent for not just understanding the needs and values of people but for capturing these and translating them into the product forms that people implicitly recognize and desire. This ability has led him to be described as one of the world's most influential designers.

Fukasawa's understanding of the human pysche is based on finding hints in the intrinsic subconscious, the intuitive rather than instinctive, a process he refers to as "Without Thought." He explains this as not just designing for emotions—which can lead to clunky products that try to fit certain human needs, but that often fail—but designing instead products that tap into the little unthinking acts that smooth out our days as we negotiate our environment and surroundings.

These are innate to all people, irrespective of culture or nationality, and he is hence able to inspire products that harmonize the relationship between people and their surroundings and which have universal, global appeal. It is often said that whilst his designs might be new to the market, people feel as though they have been acquainted with them before.

By mapping this understanding of people against different product genres and specific corporate design aspirations, he has been able to produce a canon of award-winning work that includes kitchenware, clocks, mobile phones, packaging and furniture, for a vast range of clients that include HTC, Alessi and Vitra.

2.32

2.31

Figure 2.31
Fruit juice packaging. "I imagined that if the surface of the package imitated the color and texture of the fruit skin, then the object would reproduce the feeling of the real skin."

Figure 2.32
PlusMinusZero humidifier. The PlusMinusZero humidifier brings in historic Japanese design features to turn a traditional angular metal designs into an object d'art. Fukasawa explains that he picks up on existing human behaviors and reflects them in his work, calling this "design with a subconscious mind."

Recognizing design trends

This section looks at some of the trends in design through which it is possible to see and gain inspiration from the way that other designers have been creative in the way that they are reading, defining, and responding to market needs.

Design movements

When the "prevailing inclination" of designers is said to uphold or follow a similar set of ideals over a period of time, then it's possible to see the emergence in history of design movements. These movements include, for example, the Scottish School, the Arts and Crafts movement, Art Noveau, Surrealism, American Industrial Design, Streamlining, and Modernism. With the benefit of hindsight, it's possible to consider how these ideals were shaped by the prevailing beliefs, technologies, social norms, and politics of the time and to understand how designers built these ideas into their products.

Contemporary trends

While it is relatively simple to look back in time and interpret the trends, it's harder to see current trends in part because the late twentieth-century tendencies are an eclectic mix of styles and thinking often defined by skepticism toward theories and usually referred to as postmodernism. However, it is possible to look at the broad trends of designers, products, adverts, and thinkers to draw some design conclusions.

2.33

Figure 2.33
ECKO AD-65 radio. The modernist designer Wells Coates applied crisp and uncluttered thinking to the design of clocks, radios, aircraft, and living spaces with an inclination toward the use of new materials without the hindrance of being encumbered by historical trends. The ECKO AD-65 radio used a molded Bakelite casing in a redefined shape that was both modern and cheap and highlighted the product features in radical change from the traditional wood cabinet housing or contemporary radios, making it the iPod of its day.
Image DEA/A. DAgli Orti

Individualism

As the population grows and our world becomes more congested, it could be said that people feel more of a need to be special within the massed crowd. A design reaction to this has been to provide products that are individually customizable, giving consumers a feeling of empowerment and a feeling of satisfaction by realizing their own uniqueness. This form of "pimp my product" might include anything from clip-on mobile phone covers and products with adaptable programmable open source software to products with transformation gel form coatings that change shape over time. Nike's "consumer decides" philosophy is a strategic direction taken by the company that allows customers to design their own footwear. It's driven by the acknowledgment that an eighteen-year-old

has different tastes and demands to that of a twenty-two-year-old.

This position can be taken a step further to argue that consumerism is driven by the desire to find not just individuality but actual personal happiness through the purchase of material possessions. Social commentators such as Karl Marx were noting as long ago as the nineteenth century the move away from consumption to meet peoples "needs" to consumption to meet societies "wants." An example of how this reaction translates into design is evidenced by both the range and the rapid turnover of fashions and products. Another is through the process of choosing products or brands that we feel will help to define us. Apple or Windows? Nike or Walmart? Straight leg or boot cut jeans?

Figure 2.34
Do hit chair for Droog by Marijn van der Poll. This demand for individualism is a significant design trend, but it is mocked here with Droog's typical dry humor by the Do hit chair. Designed by Marijn van der Poll, the Do hit chair is one that can be fashioned by the owner smashing, hitting, and pounding it into shape (using the sledgehammer provided with the purchase).
Photographs: Robbard/ Theuwkens (styling by Marjo Kranenborg (top) and Gerard van Hees (bottom))

Rebellion

Counterculture is another response to our society including movements that seek to oppose the economic and political frameworks that are seen to create life's stresses and strains. This takes in movements from "switch of the TV week" to "reclaim the streets" to "adbusters." It might even be considered anticool. The rebellion response within design might include generating simple, low-cost, and nonbranded products, or products that are socially and community motivated rather than purely profit orientated.

Organic

If you look at the products around you now, you may find it hard to find anything that is perfectly square or linear. If you are sitting at your desk, look at your chair, lamp, mouse, pen or the telephone; how many perfectly straight lines are incorporated in these products? The current trend for smooth, curving organic shapes is in some cases a reaction and shift away from the clean, symmetric designs of earlier years. Organic lines mirror natural forms, where few straight edges can be found, giving products a more natural and aesthetic feel to them. Computer-aided surface and solid modeling, coupled with advances in manufacturing techniques, has facilitated this trend.

Figure 2.35
Harley-Davidson Breakout. "We're not really about transportation: it's not about getting from Point A to point B. It's about living life in the way you choose." Mark-Hans Richer, CMO, Harley-Davidson

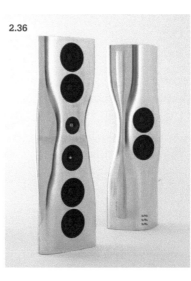

Figure 2.36
Muon speakers by Ross Lovegrove. Sometimes referred to as "captain organic," Lovegrove's work is inspired by the evolution and microbiology of the natural world. The Muon speakers are made by heating and super-forming sheets of aluminum.

Pleasure

Leisure activities have a similar function to play in allowing people to de-stress and detox. Compared to the activities of previous generations, many people are also finding different leisure activities allowing for further design scope. Hybrid mountain bikes, LED reading lights, social networking and gaming sites, parkour, and home cinema represent some of the developments in leisure style products and activities.

Play has a powerful way to relieve people of everyday stresses and strains, and some designers might aim to develop products that not only perform a function but also help people to have fun. There are numerous examples of these products including the ironic or joyful products of Alessi.

Escapism

As systems, technology and organizations take more and more control of our lives, it is easy to feel dwarfed and swamped by these "massed but invisible forces" around us. Products that provide a way of escape or that legitimize behaviors that allow forms of escape can have a role in easing this burden. The "slow movement" might be considered related to this, with an emphasis on products and behaviors that seek to slow down the frantic pace of life. Ironically, technology can also help to escape this world. For example, personal music players with earphones or headphones are ways of shutting away the noise, intrusions and dramas of the outside world, whiles the listener enjoys a period of legitimate, restful calm.

Figure 2.37
Hourglass Desk Clock (top right) by David Dear for Kikkerland and Rainbowmaker (bottom right) by David Dear for Kikkerland. A graduate of Rhode Island School of Design, David Dear is one of many designers who has made contributions to Kikkerland Design's range of products that bring witty, novel, or effective features to mundane items with an intrinsic appeal to many consumers. Dear's Hourglass Desk Clock is most unusual; the red line indicates the time, with the minutes at the top, the hours at the bottom, and the seconds indicated by the spinning disc on the top. The Rainbowmaker harnesses solar power to rotate the crystal, and in doing so it refracts sunlight and casts rainbows.

2.37

2.38

Figure 2.38
LG OLED curved TV. Organic light-emitting diodes (OLEDs) are electroluminescent compounds sandwiched by electrode film. These can be flexible as used here in LG's curved screen TV allowing a fully immersive experience.

"I think that this will be the goal of design in the future (or at least, my goal for my future in the sphere of design): transforming the gadget function ascribed to objects by the consumer's society into a transitional opportunity, namely into an opportunity for consumers to improve their perception of the world."

Alberto Alessi

Adult child

There is much written about the observation that children are trying to grow up too early to become adults, whereas adults are trying to become more childlike. The term *kidults* is often used to describe this mixed-up phenomena. The consequences on design include adult products for children. Hence, a designer might produce a miniature working versions of adult things for children rather than simplified, toylike replicas. A small metal toolkit in a toolbox rather than a plastic blow-molded set in a plastic toolbox. Educational toys have similarly become the largest growth sector in the toy market, particularly for electronic products such as miniature laptops.

Conversely, designers can produce more childlike toys and products for adults. Electronic games are also a growth area for adults, and along with novelty kitchen gadgets, amateur do-it-yourself versions of professional tools and quad bikes might all represent adult versions of toys and hobbies.

GAMIFICATION

The term *gamification* is attributed to the British programmer and inventor Nick Pelling and revolves around the principle of applying the attributes of game playing such as competition, fun, achievement, and reward to the design of product and services in order to increase the user experience.

THE "INTERNET OF THINGS"

The *Internet of Things (IoT)* is a term proposed by British engineer Kevin Ashton, which recognizes that more and more of our products are uniquely identifiable and capable of sensing, linking and communicating through internet technologies. There has accordingly been a host of rapid developments in 'smart' products that appear to think more intelligently or act more helpfully or independently.

Future Design

Futurology is a field of study dedicated to the science of evidence-based speculation and postulating of future worldviews. Some future trends are more likely to be perceived as relevant than others.

Technology

Technology is an endemic part of our world that has enriched many facets of our lives and looks likely to continue at its accelerating pace. Some people like technology-rich products because of the increased functionality, or enhanced performance that they can bring. Some people like technology-rich products for the simple reason that they are technology-rich. Quite often younger designers will be technically adept and technically motivated, but it is important, however, that they recognize that there is another side to the technology argument. Technology can, however, be stressful, controlling, and isolating. For example, some people may simply dislike the social intrusion that mobile phones bring into public and personal spaces and not buy a mobile phone at all, or dislike the notion that a toaster may be making assumptions about your everyday routines without your input.

Figure 2.39
iRobot Scooba
450. This auto-
mated floor
cleaner uses tech-
nology to remove
a chore that not
only takes up time
but also is prob-
ably considered
very dull by most
people.

Sensory design

Design is sometimes thought of as a purely visual practice but the importance of touch, smell, and feel should not be overlooked. Look around you, and you rarely see people with hands that are not doing something; holding bags, tapping pencils, and exploring parts of the body. The somasensory feedback from touch is stimulating and memorable; hence, incorporating it into products is a valuable activity. Smoothness is one aspect. Shot or sand blasting die molds gives plastic its familiar stippled texture, whereas metalized plastic has a smoother, friction reduced finish. Shape, thickness, and thermal conductivity can have an equal role in the tactile experience. Developments in gel and polymer coatings provide rich areas for designers to explore.

Sustainable design

Concerns about the environment have been raised since the 1950s when some people began to question the assumption that the earth and its natural systems could accommodate and absorb all human activity. Only recently has it been widely accepted that human activity is having a negative effect on the planet. One of the environmental effects of human activity is the depletion of the earth's natural resources; another is the generation of greenhouse gasses The production of electricity releases gases that become trapped within the earth's atmosphere. This then traps the sun's energy within our biosphere in much the same way that air in a greenhouse is warmed up. The gasses are, hence, referred to as greenhouses gasses, and the effect is warming up the planet. Different

Figure 2.40
iPhone by Apple. The iPhone is smooth and solid with a breath-taking user-centric touchscreen control. Even though its perfor-mance and cost have been the subject of some discussion, the iPhone's tactility and touch sen-sitivity alone have made it a beautiful product for many people.
Photograph: Apple Inc.

"There are professions more harmful than industrial design, but only a few of them."

Victor Papanek

weather patterns are likely to mean more extreme weather effects such as drought and storms, with catastrophic effects on animals and plant life.

The term *sustainable development* was defined by the World Commission report on Environment and Development (the 1987 Brundtland Commission) as "development that meets the needs of the present without compromising the ability of future generations to meet their own needs." Sustainability, therefore, encompasses broad concepts of economics and politics in addition to its more familiar environmental considerations. Sustainable design responses can range from using less materials, sourcing locally materials that cut down on global transportation to using lower energy production processes of creating products that encourage more eco-friendly lifestyle changes.

Environmentally friendly products have, however, been slow to appear on the market, but there is now a growing—and essential—trend toward greener products. This trend is being driven by designer realization, corporate and consumer concerns, and in some cases governmental legislation. It is being enabled by new materials, new business models and developing understanding of eco design principles such as environmental benchmarking.

2.41

Figure 2.41
Knotted Chair by Marcel Wanders. This armchair is made of macramé knotted carbon and aramid fiber cord with an epoxy resin finish. The materials might not be considered environmentally friendly, but it is a good example of achieving an environmentally friendly product through low material content.
Photograph: Cappellini Courtesy of Marcel Wanders Studio www.marcelwanders.com

Case study
Luigi Colani

Luigi Colani is a maverick industrial designer, with a walrus moustache and a proclivity for white suits and large cigars. He was born in Germany in 1928 and studied sculpture in Berlin and aerodynamics at the Sorbonne in Paris, subsequently working with or for companies such as McDonnell-Douglas, Fiat, BMW, Mazda, and Sony. His collaborations are reputed to cover over 5,000 industrial and consumer products and range from cars, planes, and trains through to computers, headphones, cosmetics, sunglasses, and uniforms. Colani and his team have won many awards and achieved many firsts for their work including the BMW 700 sports car and the Canon T90 camera. Colani is considered by many to be one of the world's most influential designers.

Colani's list of products and awards is impressive; however, it is perhaps his style for which he is best known. Rounded, organic forms that incorporate a heavy acknowledgment of nature characterize his biodynamic or biomorphic style. Colani refers to himself as a three-dimensional philosopher and asks why he should join the straying masses and make everything angular.

Colani's style has often been too challenging, too futuristic, or too grand to be accepted by everyone and his larger designs are often not taken up at all, but this is of little concern to the man himself, who is keen to impress his willingness to experiment.

2.42

Figure 2.42
Poly-COR-chair. This plastic chair draws on the single leg design of Eero Saarinen, but incorporates a cantilever feature. Although difficult to produce, this 1968 design remains relevant today.

Colani's style, approach, and ideas have numerous supporters who straddle both the eastern and western hemispheres. He has equally influenced a generation of designers with interests in anatomy, technology, and bioforms. Designers such as Ross Lovegrove and Karim Rashid might all cite Colani as a major source of inspiration. It is an influence that is seemingly gaining more advocates because even though the feats of other designers and design groups might fade over time, Colani's work seems even more relevant than ever to today's markets. Tactile materials, technology and anthropomorphism, and fluid curves enabled by modern CAD software and environmentally driven calls are all highly visible trends in today's products, but all of these issues are evident throughout Colani's fifty-year body of work.

Figure 2.43
Luxus motor home. This is a streamlined motor home concept.

Chapter summary

This chapter has explored the role of sketching, modeling, and attitude in the development and evolution of ideas. These are well-recognized design tools, but their implementation is often distorted, and this final section has aimed to show more clearly some of the purpose and pitfalls of these practices. The chapter has also concluded by exploring some of the ways that a whole raft of very good designers are reacting to different market needs. For example, some of the trends that are emerging from these designers include generating designs that help people to cope with an increasing complex environment.

Assignments

1. Debate whether the gamification of products is a good or a bad thing.

2. Designing products often involves making things faster, better, and stronger. There is, however, a school of thought that says things should be slower. For example, do you really need to make a cup of tea instantly? Do you need to stand over the kettle tooting while you wait for it to boil. Imagine a kettle that slowly brings water to the boil, saving energy as it does so and letting you know when it's ready (as kettle used to do). Try generating the concepts behind a range of "slow" products.

3. Provide a one-page business memo as to why environmentally friendly products have been slow to appear.

4. Spend a whole day in creative mode: dress differently, go to work differently, eat differently, and mix with different colleagues. Record your feelings and observations.

5. Practice problem solving using a range of different creative tools. Problems might include how to make it easier to sharpen a pencil, how to make it easier to open a carton of juice, or how to stop a small fire from spreading out of control.

6. Innovate a hair dryer or pen based on the adaptation of its shape and tactility.

7. Construct a list of everyday products and artifacts, and try to adapt these for sensible but completely different uses.

8. Consider and sketch twenty different ways to make toast other than with a conventional thin wire electric toaster.

Product development

To get to this point in the design process may have required much effort in the research and ideations stages. In the next phase, though, an idea will be turned into a tangible product and in many ways this is actually where the hard work really begins. Difficult decisions have to be made as to which concepts should be progressed further and which should be rejected. For those that are selected, much work has to be done adding the detail needed to turn a concept into a fully working and optimized reality. It is at this stage too that development costs usually start to accelerate. This chapter outlines some of the issues involved in the selection process and some of the factors to consider when working on the developing product's form and function.

Concept selection

Time and resource restrictions mean that not all concepts can normally be taken forward to development so the completion of the concept phase is usually a convergent process involving sifting through the ideas to arrive at the small number of ideas that look most likely to succeed. It's normal to select between one and three ideas, to go into the development stage. This is still a range, but it is a more manageable range that might now be pitched toward clients, managers, or investors.

The ability to pitch can be vital at this stage. Failure to demonstrate ideas well means that the hard work done so far can be lost, or that weaker ideas will be taken forward. Months or even years of effort can boil down to a few moments of time. Being able to get ideas over clearly, concisely, and persuasively is, therefore, a key design skill.

Figure 3.1
Slice box cutter. The curved carbiner grip adds comfort and hooks easily into belts for quick access, easily seen with a distinctive orange flash.

3.1

"What it boils down to is one per cent inspiration and ninety-nine per cent perspiration."
Thomas Edison

Decision making

Many people might see design as a purely creative function; **decision making**, however, is an example of the type of analytical and critical thinking needed throughout the design process. Decisions are constantly required in design: how to go about appropriate research, which user groups to research, how much to spend, where to start detailed design, which CAD packages to use, what materials to use, and so on. It's also clearly required when deciding which ideas to take further forward.

Hence, decision making is an often-overlooked core design skill. In large organizations, where a number of different projects may be underway at any one time, and where a number of people may be needed to convene to make decisions, it is normal to establish a series of key points referred to as phase or stage gates where projects can be evaluated.

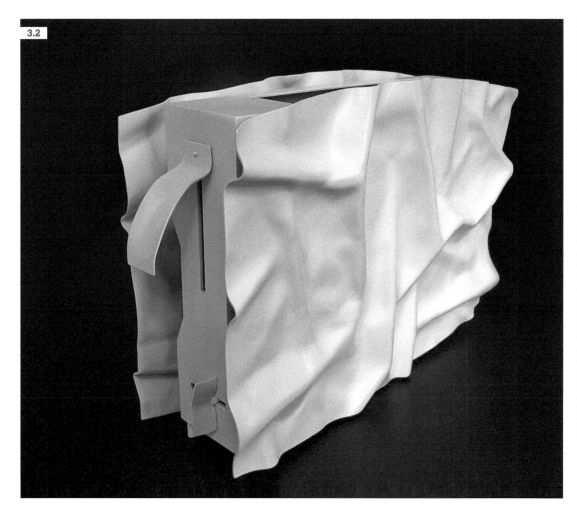

3.2

Figure 3.2
Hotpaper toaster by Olivier Grégoire. There might be a range of areas to explore in the design of a toaster including, for example, ways of toasting bread, methods for controlling the function of retrieving toast after it has been cooked, and the visual aspect of the design. The Hotpaper toaster explores ideas around visualization. Of his design, Grégoire said, "I wanted to create something more unique by a contemporary approach of material. As for the design, I wanted something that disregarded the product itself by favoring the visual impact."

QUALITY FUNCTION DEPLOYMENT

Quality function deployment (QFD) is a slightly more sophisticated decision matrix technique that has arisen from within Japanese industry. It maps customer requirements (typically taken from the criteria in the user brief) against the features created by the designer using a scoring system. In this way it's possible to quantify (i.e., measure) how well a designer has done in meeting the needs of the consumer. It includes other features, too, such as assessing technical complexity or benchmarking against rival designs. It's a visual technique so is a powerful communication tool for design teams.

Quality

Designers can sometimes be biased in their decision making. This might be because they would prefer to work on ideas that are of personal interest, or because the ideas are more exciting or challenging than the ideas that would best suit the needs of the market. Even though most people might perceive "quality" as a measure of luxury or high cost, in the design industry the term *quality* has come to define the principle of meeting customer needs. An expensive sports car might for example not be a quality product if the consumer needs are to transport a family of small children to school every morning. Decision matrices are useful tools for achieving quality by measuring concepts against criteria and generating unbiased quantitative decision data based on evidence rather than personal preferences.

3.3

"Quality has to be caused, not controlled."

Phil Crosby

Figure 3.3
Home Hero fire extinguisher. The design of the Home Hero meets many consumer needs such as single-handed ergonomic use, with a house-friendly sleek, softly colored form that is easy to clean. By incorporating smoke and carbon monoxide detection, it may surprise and surpass consumer expectations.

3.4

Figure 3.4
Airbus A380.
The A380 is the world's largest passenger airliner. However, it is a high-risk product that has required €8 billion investment with no guarantee of any return. Why risk basing decisions on an interior styling to small focus groups that may not be representative of the bigger market and that may lead to outcomes that the majority of people dislike. Online forums allowed Airbus to canvas opinion on their design from anybody with access to the Internet. Airbus also ran a "fly your idea" contest allowing engineering teams a chance not just to comment on designs but also to actually participate in the project.

"It's very easy to be different, but very difficult to be better."
Jonathan Ive

User feedback

In the time taken to research problems and generate ideas, it is possible that user requirements may have changed. It's also possible that designers have not interpreted needs or developed solutions in quite the right way. Validating ideas is the process of taking concepts back to the market to test whether the ideas being developed align with the intended user's needs. This can be done through traditional methods such as focus groups, which use small groups of people convened to discuss or try out new products or through the Internet which enables much larger numbers of opinions to be garnered from a wider range of people through crowdsourcing and crowd voting.

Figure 3.5
Bagetty microwave by Martin Zampach. This microwave concept was designed to meet the market for cylindrical fast food products such as hot dogs and bread sticks.

3.5

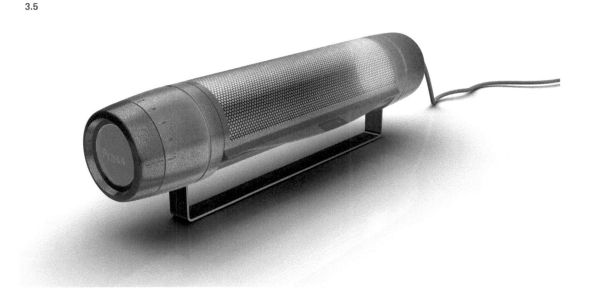

Product specification

A product or design specification draws together and summarizes thoughts, research, imagination, performance targets, and data and turns the outline brief into a more detailed statement, which includes the proposed solution. This should not be confused with a technical specification, which is a description of the completed product and its characteristics. The specification should define what the completed product should aim to do. As such, it provides clarity for everybody involved in the future development. It also acts as a catalyst and a reality check between all stakeholders. It might not only clarify what is known as being required but also outline what is not known and needs further work. Therefore, it can be an evolving document.

It is important to get it right. After the more detailed development process starts, stopping or changing your mind becomes more difficult. One reason for this is that costs start to mount up making it harder to admit mistakes. Another reason is that design usually involves groups of people that include other designers, purchasing personnel, production workers, salespeople, clients, and consumers, and it can be very hard to get a group of people to all agree to change direction.

3.6

Figure 3.6
ResQtec hydraulic rescue tool. Equipment designed to extricate trapped people must be designed to act quickly and powerfully, yet it must be lightweight, comfortable, and safe for the equipment operator. This design has included bold colors and polished surfaces to help enhance the product's visual appeal.

Meeting the specification can also be expensive because much of the development costs lie in experimenting, researching, prototyping, and manufacturing, all of which contribute the largest portion of any new design. Once time and money have been invested, it becomes harder to go back. The product will then inevitably be released—flawed. The lack of a complete and thorough written specification is, hence, one of the main reasons cited behind product failure. Get it wrong, and you are committing time, money, and effort to a poor or doomed product.

**Figure 3.7
StyleiD by
Philips.** LED lights have revolutionized low-energy lighting, but issues of heat, color, and light quality require significant design attention alongside the need to make products that look good and are easy to manufacture, install, and operate.

3.7

Detail design

Having created a specification to clarify the exact design requirements, it is important to think through where to start the process of fleshing out the design details. Which criteria, which features, or which functions do you look at first? What can make this even more challenging is that many criteria can be contradictory. Making something more durable to make it last longer will conflict, for example, with criteria calling for minimal weight or minimal cost.

Given that different starting points can significantly affect the ease of the detail design process and the nature of the product that emerges, it is surprising how often this aspect of ordering can get overlooked. A lot of research into design methodology revolves around answering this question of knowing just where to start. **Design for X**, for example, involved the development of a set of guidelines about the products criteria and, importantly, their links. Axiomatic design uses a matrix to map the key issues and a set of design principles (axioms) to facilitate the ensuing process.

FORM FOLLOWS FUNCTION?

The principle of form follows function was developed early in the twentieth century and was taken up by the Bauhaus and industrial designers such as Loewy, Dreyfuss, and Papanek. If you are designing a wine glass, for example, and focus on the form, generating a fantastically evocative flute, you might end up with something that holds no liquid and can't easily be drunk from. If you focus on the function, holding 33 centiliters with a thin rim 120 millimeters long, you might end up with a wine glass that looks like any other (opponents arguing that if the function is optimized, then the form becomes ubiquitous—simplified to the point where one design fits all, are boring or impractical). In reality, the type of product is important in this debate, and issues such as meaning and user experience have risen to take as much prominence in thinking as form and function alone. However, it can be informative to approach a new design from these two purist perspectives.

"Design is the fundamental soul of a human-made creation that ends up expressing itself in successive outer layers of the product or service."

Steve Jobs

Case study
Royal Philips

Royal Philips of the Netherlands employs over 100,000 people in industrial sectors that range from medical and healthcare through to consumer electronics, grooming, and lighting. For many years Philips has been known as a technology pioneer and it can count around 59,000 patent rights in its portfolio generated from its research bases in Holland, India, and Shanghai (China) or from its numerous working partnerships.

The Phillips group has more recently transformed its brand focus to emphasize more consumer lifestyle and user experience, using both experience design and technological research in its drive for meaningful innovation.

This mix of technology and user is evident in products emanating from its lighting division, which has evolved since its technology driven on/off light bulb design of 120 year ago to explore new technologies including LEDS and OLEDs mixed with user-centric themes of user/light relationships, expression, communication and life. For example, mood, color, environment, energy, simplicity, inspiration, range, connected products, and identity all need to be considered within the developing products. Philips might refer to these as solutions that use "light to enhance life" and that go beyond the physical product to include the entire user experience, for example, enabling online reviews, associated apps, webinars, twitter, and even knowledge-based instructional courses.

The company uses "design thinking," involving teams and tools such as "experience flow," to help develop these products successfully, led by people labeled as thought leaders having the ability to inspire and make things happen. Future products developments from these teams are likely to include lighting incorporated into fabrics and materials, and lights that are used to ease pain, boost health, or revolutionize the way we interact with other things. Understanding the "light recipe"—the mix of intensity, hue, and duration, for example—and the way to implement this recipe into different environments at different times represents the next challenge for designers developing the next generation of products at the company.

3.8

Figure 3.8
To Be Touched. This lighting controller is designed to be intuitive and easily interactive.

3.9

Figure 3.9
FreeStreet. This minimalist
street lighting system uses
LEDs to save energy and
remove the intrusiveness
of traditional street lighting
instillations.

3.10

Figure 3.10
Style iD light. Elegant and low-cost
modular lighting that gives consumers
options over optics, size, output and
accessories with flexibility in the configura-
tion of the lighting.

Functionality

Adding in the functionality that brings a product to life can simply mean choosing the right materials or manufacturing process, but it is likely to also include optimizing the way it operates. This means not just its mechanisms and performance but also the way it interacts with users. This interface between people and product is referred to as **ergonomics** or human factors (or interaction design for the interface between digital interfaces and people). Some designers argue that there are differences between the terms, but they are essentially synonymous and require an understanding of issues such as biomechanics, psychology, and the contextual environment in which users and products operate.

"I just want things to work properly."

James Dyson

Engineering

Engineering includes mechanical, electrical, and electronic engineering and is concerned with using knowledge to resolve problems and with the development of machines and tools. It lies at the heart of making products work well—designed so they don't break and perform to their optimum. Some designers might know a lot about engineering, but others might know very little, but at least some knowledge is required because engineering is at the heart of making products work well. The minimum amount of knowledge is the understanding of basic engineering principles and terms that allow designers and engineering to communicate and designers to comprehend the engineering outcomes.

One of the most effective and cheapest ways of engineering a working design is by using mathematical models. These can be used to locate pivot points, to calculate sizes and outputs, or to predict how designs will behave when variables such as force or temperature are changed. The model need not be complex. Simply understanding that force = mass x acceleration or other formula can save many hours and wasted effort with more elaborate experimental approaches. Mathematical models are, however, only symbolic representations of reality, and answers should be treated as informative rather than exact.

3.11

Figure 3.11
Arduino D7756M. A number of programmable logic controllers have become available to help non-electronics experts to create working electronic models. These are effectively low-cost mini-computers' that can be programmed using simple language. An Arduino is an example of a simple microcontroller that can be used to develop projects that have interactions between sensors, switches, motors and lights for example. It links easily to a personal computer, and as an open source (i.e. available to everyone to work on) programmed has lots of online community support.

Experiment

A designer might be asked to design anything from a pump, to a spoon, a washing machine, a vase, or a coat. It is impossible even for an expert to know everything about anything (or anything about everything)! Sometimes reading round a subject might lead to enough knowledge, but at other times experimentation or calculation is necessary. Advanced products, for example, might go beyond what is more commonly known and into the realm of physics and new science. For example, plastic electronics (electronics made on carbon such as graphene rather than silicon) is promising a whole new branch of products such as roll-up screens or illuminated ceilings. Another core designer skill is then being able to engage with the experimental process and practice.

There are rigid protocols around experimental techniques and a whole field of study around the Design of Experiments (DoE) aimed at ensuring that the outcomes from research and experiments are robust and valid. Issues include for example controls, sample sizes, randomness and bias. Where people are involved then issues of safety and ethics become pivotal. The cost of this type of work can be expensive and statistical techniques that can shortcut findings are popular, including the Taguchi method after a technique popularized by Japanese engineer Genichi Taguchi. The technique enables the number of experiments to be minimized whilst still identifying the key factors needed to be optimized in a design, allowing the designer to concentrate on these rather than lesser functions.

3.12

Figure 3.12
Canon EOS 70D camera. Most manufacturers have developed clever solutions to reduce the problem of dust on the image sensors of digital cameras with removal lenses including piezo crystals to ultrasonically vibrate the filter surface of the image sensor. The Canon camera is regarded as particularly innovative as each pixel has dual function capability across most of the sensor allowing for a very effective live autofocus capability.

"Good design is probably 98% common sense. Above all, an object must function well and efficiently—and getting that part right requires a good deal of time and attention."

Terence Conran

Wood

When it comes to selecting materials, wood is well known as one of mankind's earliest and oldest design materials. It comes in a wide variety of forms with a wide range of properties. At the end of its working life, it decays naturally or can be burned to recoup heat energy. The carbon dioxide emitted is considered equal to the carbon dioxide that is absorbed during its growth process, making it a carbon-neutral material. It can be sourced locally or from sustainable sources authenticated to replace trees as they are felled.

One of the problems with the use of timber, however, is the presence of weaknesses such as knots and fissures, which make it impossible to guarantee its consistency or integrity in mass production. Wood-based products that work by reforming fibers of particles overcome this issue and include medium-density fiberboard (MDF), melamine-faced chipboard, and plywood, plus an increasing number of new products including wood-plastic composites, which can, for example, be molded and which provide designers with new opportunities. The disadvantages are the chemical nature of the glues and resins used as binders.

3.13

Figure 3.13
BSG WOOD.b bike. French company BSG has combined plywood with steel to create an elegant but practical and versatile bike. A choice of woods including beech, ash, and rosewood allows for customer choice.

Ceramics

Ceramics are nonorganic, nonmetallic materials such as concrete, clay, brick, and glass that can be molded or sintered into shape. They are generally hard, porous, and brittle making them ideal for low-volume tableware, lifestyle products, or for specific applications where the electromechanical properties of ceramics can be best suited. These properties can be enhanced with oxide and nonoxide inclusions. For example, titanium carbide is used in scratchproof watches and zirconium dioxide in ceramic knife blades that are harder than steel and stay sharper up to ten times longer.

3.15

Figure 3.15
Umbrella Pot designed by Kouichi Okamoto, Kyouei design. The pure white lines of this ceramic umbrella pot give it a clean, stylish, and contemporary look, although it is made from *Tokoname-yaki*, a kind of traditional Japanese ceramic of clay and feldspar (or silica) glaze that dates back to the 1100s. Water that drains from the umbrellas placed within it help to provide a drink for the plant through internal holes inside the design.

3.14

Figure 3.14
Chaise Lounge No.4 by Tom Raffield. Raffield's design incorporates local, unseasoned timbers such as oak and ash and uses innovative steam-bending techniques to create complex, expressive, and challenging products. In this example, the product represents a craft-based approach to design.

Metals

Each new development of metal—bronze, iron, cast iron, and steel—has spurred civilization forward. The major metals used today are iron, aluminum, copper, zinc, and magnesium. These are hard, strong, dense materials that can be molded, cut, or formed into shapes.

They are also often combined with other elements to form **alloy** materials with a wide variety of characteristics such as impact resistance or tensile strength. Aluminum, for example, can be found in cars, bicycles, saucepans, watches, and cans but rarely in its pure form, which is relatively soft and usually includes elements of copper, zinc, magnesium, manganese, or silicon to improve its performance characteristics.

Metals can be recycled. For example, in the United Kingdom, roughly 40 percent of steel is comprised of recycled metal. In recent years, the price of some metals has started to increase as they become harder to extract from the ground, increasing this emphasis on recycling.

Polymers

Plastic is a term usually used to describe oil-derived materials. Because *plastic* is also a term to describe the malleability of materials including metals, the more correct way to describe plastics is *polymers*, which defines materials that have repeating chains of molecules. This includes both synthetic and natural materials. As a design material, there is a wide variety of types with properties such as flexibility and color that can be adjusted with ease and formed into complex shapes and sizes.

Developments in polymer technology include high-performance engineering plastics that can act as straight replacements for metal, whereas smart materials have properties that can be changed through external stimuli such as impact, stress, and temperature giving the designer unique behavioral opportunities. For example, shape memory polymers can be deformed into one shape but induced to form another through a temperature change. Chromogenic polymers

Figure 3.16
Aluminum bottles by Sigg. Sigg water bottles are made from a single piece of aluminum that is impact extruded. This means they can be lightweight but do not have a seam, thus ensuring a constant aesthetic and durability. The inside is hygienically coated to help alleviate and possible contamination of the of contents.

3.16

change color through an external stimulus such as light-sensitive sunglasses.

The drawbacks of polymers include degradation over time and difficulty in recycling for some materials—particular those polymers grouped as thermosets—although advances in technology are making this easier.

Material properties

Most designers should know something about material such as how different material types can suit different purposes. Knowing more about the basic properties of materials will, however, allow a better understanding behind the selection and use of materials. The most common properties are mechanical and include density, hardness, strength, toughness, and ductility, but other characteristics such as electrical, thermal, chemical, magnetic, and optical properties might need to be considered for specific product functions.

These properties have specific engineering meanings. For example, tensile strength is the maximum stress (i.e., the force applied on a specific area) that a material can withstand. Brittle materials like glass might simply reach that limit and then snap, but most materials including metals will stretch a little first, returning to their original shape, up to a point (the yield point) where they deform permanently and ultimately break (the ultimate tensile strength or UTS). The property of a ductile material to stretch (deform plastically without fracturing) is useful in allowing products some "give" when under pressure rather than immediately snapping and in allowing materials to be manufactured by forming processes. This pattern of material behavior under stress can be mapped through a Young's modulus diagram.

Materials have properties with defined values that can be used to calculate, for example, how thick or thin a part should be to avoid failure, using stress and strain calculations. **Finite element analysis (FEA)** uses computers to calculate stress factors in products or components. It does this by dividing complex shapes into a mesh of much smaller elements and then using the power and speed of its processors to calculate the numerous algebraic equations that link and resolve each mesh. This can be a powerful tool for optimizing the performance of products that can be easy to do through bolt-ons for CAD packages. If applied poorly though, without a background understanding of the principles of stress or simulations, it can easily lead to failures in scaling or accuracy. Even though most properties are constant, they can change according to temperature or pressure, and fatigue failure or creep can cause failures at levels below those expected.

BIO-BASED POLYMERS

The development of polymers made from biomass rather than fossil fuel sources offers some hope for more environmentally friendly plastic materials. These bio-based polymers are made from plant constituents such as starch, cellulose, or polylactic acid and are derived from sources such as corn or sugar.

New materials

Developments in new materials are being made continually forging new boundaries in material properties and behavior. Gels and foams, including foamed metals, all offer designers exciting design opportunities. Nanotechnology, for example, is concerned with manipulating atoms and molecules on a scale of 100 nanometers (about 1/1,000 millimeter) or smaller to make new materials or devices within that size range. Applications include bespoke molecules in sunscreens to block out harmful rays, small motors and fans inside electronic circuits, and carbon nanotubes that have mechanical properties that are many thousand times greater than that of steel.

Composites are combinations of materials that can occur naturally but can also be explored to generate new materials with new properties. Aramid fibers used in combination with carbon and glass fibers and an epoxy resin for example are light and strong, and the trade marked example of Kevlar is five times stronger than steel. These composites are used race cars, helicopter rotor blades, sporting products,

Figure 3.17
MT3 rocking chair by Ron Arad for Driade. The MT3 rocking chair is made from rotationally molded polythene.

3.17

car tires and body armor. GLAss-REinforced Fiber Metal Laminate (GLARE) is another type of composite, a material composed of very thin layers of metal such as aluminum interspersed with layers of glass-fiber bonded with an epoxy matrix. Although it is a composite with lay-up options that can be matched against stress points, it is constructed using conventional metalwork techniques and is used, for example, in the manufacture of the Airbus A380.

It can be hard to keep up to date with these material developments, but the designer must try. Journals, exhibitions, networks, and technology transfer agents can help. There must, however, be a subsequent evaluation of weighing the suitability of these materials against cost and against the risk involved in using technologies whose long-term effects may yet be unproven. For example, nanotubes are of a similar size to asbestos particles, and there is a fear that these may have a similar disruptive influence on the body.

TOXICITY

Some materials and chemicals should be avoided because their effects, or their production, might be unknown or hazardous. For example, there are legal requirements in the use of products such as the Restriction of Hazardous Substances Directive, or RoHS, 2002/95/EC within the European Union, or the Toxic Substances Control Act within the United States.

3.18

Figure 3.18
Moscardino spork. The spork (spoon/fork combination) is an example of a product made from bio-plastic. These plastics are made from combinations of starch extracted from natural crops such as corn, cellulose, and vegetable oils and are biodegradable and compostable.

Case study
d3o

Richard Palmer studied for a degree in mechanical engineering and followed this with a course in design at the Royal College of Art in London. This unusual background took him originally to DuPont and then to his own innovation design consultancy. He is now is a founder and chief executive officer of d3o, a company based on the south coast of England.

d3o is one of a new range of smart materials. Under normal conditions, the molecular structure of d3o is loose with weak molecular bonds giving the material a soft, flexible, and fluid nature. When it is hit, however, the molecules lock together to momentarily form a more rigid material. During this locking process, energy is absorbed and dissipated rather than transmitted back out of the material. It may release half as much impact energy as EVA (ethyl vinyl acetate), another material with a high reputation for shock absorbance. This locking process takes place in less than a 1,000th of a second and the faster the impact, the faster the molecules react. This type of smart material is referred to as a "dilatant" material, meaning that its viscosity increases with the rate of shear (also referred to as non-Newtonian or a shear thickening fluid). In spite of its capabilities, d3o remains lightweight throughout the transformation, and this combination of properties makes it ideal for a wide variety of protective applications such as keeping high-value goods (e.g., iPods) safe, or as protective clothing and equipment for stunt work and urban safety or sports such as skiing, mountaineering, or biking.

The company usually works with companies to help them develop their own products. Even, and perhaps especially, with revolutionary innovations, it can require good presentation skills and examples to persuade partners to try something new. Richard notes that "Sometimes it's hard to convince people what a truly amazing innovation this is until you demonstrate it. I was wearing a prototype shirt incorporating d3o, and at one point I stood up and slammed my elbow onto the table as hard as I could, sending coffee cups flying. Once they saw me doing that—without flinching—they understood what I was saying."

Once companies buy in, the resulting products are a good example of teamwork in action. Richard is keen to stress the importance of analytical and creative backgrounds in the generation of concepts, both in his own background as well as that of his design team. Where possible, the company also ensures that the vivid orange color of its d3o material is visible in the design so that its trademark brand becomes more and more embedded in the world of high-performance products.

Figure 3.19
d3o contour sheet material. The contour sheet material is designed to "densify" when flexed, providing extra absorbance, for example, when bending an elbow or knee.

Figure 3.21
d3o insoles. Designed in association with Rightside Orthotics, the combination of impact absorption and geometric support is designed to aid performance and reduce the type of foot, knee and hip injuries caused through active walking and running impacts.

Figure 3.20
d3o trust kneepad. Internal and external kneepad designs can be used to prevent shock and trauma injuries in sport and industry.

Figure 3.22
Rival d3o Intelli-shock Pro Training Headgear. The first boxing head guard to use d3o technology helping to provide the comfort and impact protection in a high impact sport.

Form

Consumers normally see products before they buy or use them so appearance has an immediate visceral impact. That form must serve a number of other functions: protecting components, resisting damage, enabling use, and maintaining the beauty and appeal through the lifetime of the product. For some designers, product form is the essence of their type of design.

Aesthetics

The term **aesthetics** is usually used to describe the shape and beauty of a product, but it is only superficially connected with style and appearance and more profoundly rooted in perception and philosophy. For example, **modernism**, as practiced by the Bauhaus, represents a rejection of history and the frills of earlier movements to reshape the world with the aid of abstraction, scientific knowledge, technology, or practical experimentation. Modernist designs have been clean, symmetric, and functional, and it is this mode of thought that has shaped much of the world around us. Products that retain smoother, straighter, and cleaner lines often have a newer, more technical essence.

A debate exists as to whether we are currently still in a modernist epoch, or are now postmodern. It is argued that postmodernists have developed a mode of thought with more critical reasoning than modernists. There are cultural, intellectual, or artistic states lacking a clear central hierarchy or organizing principle and embodying extreme complexity, contradiction, ambiguity, diversity, and interconnectedness or interreferentiality. Products may often be playful, ironic, or simply opposed to modernism in any way. This effect might include curves more sympathetic toward naturalistic shapes including the golden proportion section that seem to find a more intrinsic harmony.

3.23

Figure 3.23
First by Michele De Lucchi for Memphis. The Memphis group was a collective of Italian designers and architects that evolved from a meeting set up by designer Ettore Sottsass in the 1980s to challenge the then current status of design. The group contested the notion that good design revolved around slick, black, humorless, and conformist products and released a series of works that explored shapes, colors, textures, and patterns. The First chair was designed from enameled wood and metal.
Photograph: Studio Azzurro, Mitumasa Fujituka www.memphis-milano.com

"'The past is not dead, it is living in us, and will be alive in the future which we are now helping to make."

William Morris

NURBS

The trend toward organic forms has been enabled by solid modeling in computer-aided design using NURBS (nonuniform rational basis splines). NURBS are mathematical models inside computer graphics and design packages that allow flexible, free form curves and surfaces to be generated easily and effectively using control points. French engineers Bezier and de Casteljau pioneered these techniques as seen in the curved shapes of Renault and Citroen cars of the 1960s.

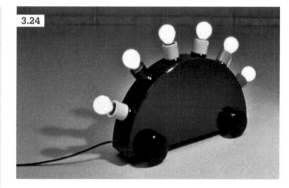

Figure 3.24
Super Lamp by Martine Bedin for Memphis. The iconic Super lamp was made from painted and lacquered metal, a typical Memphis design, and based on Bedin's early anthropometric ideas that furniture could be "like friends."
Photograph: Aldo Ballo, Guido Cegani, Peter Olgivie www.memphis-milano.com

Figure 3.25
The iMac by Apple. The original iMac was introduced in 1998, and its curved form, translucent material, and distinctive colors stood it in stark contrast to the beige boxes that characterized most personal computers at that time. It was received by consumers and designers with critical acclaim; in fact, the effect has gone beyond the computer industry with companies from different industries inspired to innovate their own Apple-inspired classics. The current desktop iMac has moved away from these earlier designs but maintains its aesthetic appeal through an elegance and thinness that conceals the technological workings.
Photograph: Apple, Inc.

Ergonomics

Ergonomics or human factors engineering are the related names given to the process of designing according to human needs in order to optimize human well-being and overall product performance. Traditionally, this has referred to ease of use, such as the physical fit between a product and its user based on **anthropometric** data (e.g., making the diameter of a hand grip a size that fits comfortably for a defined range of the population).

More modern perspectives, however, include cognitive ergonomics embracing aspects of interactions, aesthetics, expectation, perception, and sensory satisfaction. In other words, a cylindrical hand grip might be the right diameter, but there is a user expectation as to what it feels like to touch and grip and how it moves and works. A warm, shape-changing foam rubber coating might provide an improved experience compared to a straightforward metal tube.

Figure 3.26
OXO Good Grips utility knife by Smart Design. Smart Design has been awarded the Universal Design Award and the "best of the best" red dot product design award for the OXO Good Grips. The design, which includes an integrated lever, blade storage, and blade-loading mechanism, is noted for being intuitive, safe, and easy to use.

3.26

"Objects do not have meaning. But if an object is thoughtful we project meaning onto it in daily life."
Karim Rashid

Affordance

An important transition in design thinking is to view a product as not simply something with form and function but as something that can impart meaning to people. *Affordance* is a term coined by psychologist J.J. Gibson referring to the relationship between people and other people or people and things. In design, this term has come to mean the relationship (either real or perceived) between a user and a product. More specifically, the more tangible affordances of a product (or part of a product) should indicate to the user how it works or should be used. Therefore, a product should announce what it is, what it is for, and how it can be used and provide feedback that it is being used correctly. For example, a dished or concaved surface rather than a flat surface might reinforce the "push" requirement of an on/off button.

Affective design is a branch of ergonomic thinking that is concerned with the emotional effect on a person that is generated through their interaction with a product. It is the way that a product *affects* a person resulting in an emotional or behavioral response and can add significant depth to a design. For example, the aim is to deliver products that delight. Consider, for example, a table lamp that adjusts its own level of brightness to the ambient light level. The form and function remain the same, and the ergonomic action of switching on the lamp remains the same. However, there may be an emotional pleasure to be derived from having a uniform level of light while reading or from knowing that energy consumption is being optimized.

3.27

Figure 3.27
Zippo lighter.
Windproof and lifetime guaranteed, the Zippo lighter gives the warmth of reassurance to consumers with an intuitive, weighted flip lid.

EMOTIONALLY DURABLE DESIGN

Emotionally durable design, a term coined by Professor Jonathan Chapman, describes the complex factors that determine the value, meaning, and endurance of a product (or other artifact). It links the principles of affective design with sustainability, which means in simple terms that if a person has a strong enough emotional attachment to a product, then they are less likely to throw it away.

Semiotics

Semiotics is defined in simple terms as the study of signs but, in a broader product context, has come to include the study of subjects such as art, literature, anthropology, psychology and sociology as ways to explain meaning through the language of signs, symbols, myths, rituals metaphors and values. It might be thought of as the language of affordance.

Imparting meaning might mean that fast cars should look sleek and streamlined, whereas an off-road car should look rugged and robust, but there can be significant deeper subtleties at work beyond styling. For example, the Aston Martin DBS has a small, glass briquette that slides into a slot on the steel dashboard that glows deep red when the engine is ready to fire, a highly fitting function to power up a 6-liter V12 luxury sports car.

In exploring semiotics, it's important to recognize that meaning varies from culture to culture and person to person. Different individuals and societies may attach different meanings to the messages imparted. For example, black may mean cool and hip in some countries but death and mourning in others. Designers may have to consider a meaning that is optimized for a wider global audience or evaluate the maufacture of a range of designs to meet different cultures.

This can open up a fierce debate about globalization. Is it good to have products circulating worldwide for the benefit of all? It is argued that the exchange of cultures and products through globalization has helped to alleviate war and need—the McDonalds and Dell principles—suggesting that there are fewer or no wars between countries supporting these corporations. On the other hand, does it end up with omnipresent design contributing to a common world culture at the expense of individual local cultures and traditions?

"The most exciting opportunity for innovation lies in combining the knowledge systems, tools, and social and territorial assets of South and North."

John Thackara

Inclusive design

Inclusive design, universal design, or design for all are broadly similar terms for the practice sometimes incorrectly thought of as designing for the aged or less able but more correctly referring to the principle of designing products that are accessible to the widest possible audience. This could pertain to people with disabilities, but could equally pertain to gender demographics or other cultural groupings. Inclusive Design includes considerations of issues such as equitability, simplicity, flexibility, intuitiveness, and ease and low effort of use. For example, design features might include comfortable handles, easy visual operational signs, strongly differentiated colors, control through feel as well as sight, and appropriate product sizes and weights.

There are ethical reasons for making products that all sectors of society can engage with more easily, but there are economic reasons, too. Improvements in medicine and welfare mean that there is, for example, a growing proportion of elderly or disabled people in society. Moreover, inclusive design incorporates many areas of good design practice in aesthetics and ergonomics for the benefit of all.

3.28

Figure 3.28
Antique fridge by Meneghini.
This top-of-the-range fridge captures and builds the semantic values and meaning of antique styling into a modern appliance.

Case study
Matthew White

Matthew White studied mechanical engineering before taking a course in industrial design engineering at the Royal College of Art, London, and he now runs his own design practice in the south of England. Matthew is a good example of a user-centered designer who is capable of exploring all facets of a project, questioning many of the normal assumptions and traditions, and producing products that make people look back and ask why these ideas weren't thought of before.

3.29

Figure 3.29
Sandbug. Matthew White was aided by the Helen Hamlyn Research Centre (HHRC) on this project with B&Q to design power tools that could be used by older people with reduced grip. Users guide the Sandbug sander by palm rather than having to hold a handle.

Early in his career, B&Q (Europe's largest hardware chain store) asked Matthew to help the company in its quest to incorporate more inclusive products in its range. To better understand the needs of the do-it-yourself market, Matthew undertook a range of techniques that might best be described as action research. Action research is a form of problem solving that is based on a principle of oscillating between doing and reflecting to reach the optimum solution. It can be done individually or in groups and is a good technique to uncover new perspectives on the way that people and products interact. A competitor product audit was undertaken first, followed by interviews conducted with B&Q customers and staff. As ideas were generated, two-hour long focus groups composed of five people each were conducted, and prototypes were often used to help obtain both quantitative and qualitative feedback. Further feedback was obtained through user testing, a process involving the assessment of users as they performed specific tasks with various tools, test concepts, and prototypes. These were again discussed in informal focus groups. Over a period of eight months, nine users were selected and given a range of tools and tasks so that they could evaluate and provide feedback on both existing tools and new ideas. Key issues from the action research were then written into the product brief.

Matthew's subsequent ideas included separating the heavy batteries from power tools so that these could be carried independently, making the drill lighter so it could be handled easily by the less physically able. He questioned why it was that screwdrivers were invariably long and thin, a shape that can be particularly hard for elderly people to hold. White's screwdriver, the Gofer, is designed in a pebble-like shape: smooth and rounded so that it fits into a hand more easily. It also has an electric drive to aid one-handed screwing. Unlike most electric screwdrivers, it is designed with a push-and-go mechanism instead of a button. White also redesigned a sanding device so that it would be easier to hold. The Sandbug features a large domed body that fits into a cupped hand and has a generous hand strap for extra grip (the inspiration for these design details came from observing people using a horse brush). The Gofer and the Sandbug do not resemble traditional tools, but, like many inclusive designs, their features appeal to a much wider audience than just the elderly or less able, making them highly successful products.

An interview with Matthew White

Innovation always seems so easy—with the benefit of hindsight! How do you go about generating ideas for new products?

I look at the state of the current products (if they exist) and often try them and take them apart. Then I sketch and think a lot. I try out ideas simply and go from there. I try to really think about the actual use of a product and then I am inspired to give the end user the product they need. You are very strong in determining what people need from a product.

How do you ensure that your own prejudices don't interfere with the way concepts are developed?

They always do—let them! The best thing to do is be prejudiced in the right way—to think like people who struggle with products or who aren't being catered for yet. But I still fall into the trap of designing for myself. To avoid this, you have to interact with users as much as possible during development—even if just a friend or colleague.

Have you changed the way that you do your research, either over time or for different products?

Yes—it varies a lot depending on budget. The Sandbug and Gofer were all about research with those less able. It's not always practical to do a lot, but important to do it once and carry those experiences with you.

Any ideas how we might define cool [admired]?

Yes—in many ways "cool" just means it's the way it should be—or rather feels like it should be. It doesn't stand out as poor design or a try-hard!

Which other designers do you admire and draw inspiration from?

Difficult. I don't actually spend much time looking at others' work. I think you can try to just emulate them too much. Names that stand out for me are Stark and Lovegrove but that's quite obvious! I tend to let individual objects and pieces of design inspire me—the Bird's Nest stadium [in China] really has recently.

Is it easy to sell new ideas to the client companies you work for?

It's easy to get interest and some enthusiasm but far harder to keep it over the long term of product development. As budget and time increases, the sell becomes harder and harder. There is a real hunger for innovative products, though, more so than for beautiful products.

Figure 3.30
Gofer screwdriver. By providing new products such as the Gofer screwdriver, B&Q can score an advantage over its DIY store competitors and achieve a new innovation-related brand image. *Innovation* and *user-friendly* become the key words to describe the Gofer's design.

Chapter summary

Sometimes a new product just seems to look right, feel right, or work right in such a way that consumers might think that design is easy. Some products can be so good as to simply go unnoticed. Behind this is the skill of the designer who is able to understand just what a product needs and to find the ways to provide it.

The sections in this chapter have outlined some of these processes at work starting with choosing the right concept to take forward for further development. Included in this are the notions of quality, user validation, and specifications.

The subsequent development involves filling in the design details requiring consideration as to where to start plus a wide variety of knowledge relating to the way the product looks and the way it performs. New materials, engineering, and physics offer opportunities to transform even the most basic of product redevelopments into something new and exciting. The development of thinking around semiotics is also providing designers with the capacity to generate innovative products beyond simple function and form. Consequently, many good products exist today because of the attention paid to user experience.

Assignments

1. Explore a number of contemporary design groups or companies and assess whether their design work might be considered traditional, modernist, or postmodern.

2. Select a range of products and appraise where you think the designer might have started when undertaking the development.

3. Design a flashlight first emphasizing the principle of form and then emphasizing the principle of function.

4. Discuss whether bio-based polymers really are as green as they seem.

5. Produce a list of engineering terms, principles, and equations that you think all designers should have in their toolbox of knowledge.

6. Develop a two-minute pitch presentation around a product idea of your own choosing. Gather in groups of four to present your pitch and sell your idea. Vote for the presentation that best sells the idea. Discuss why.

7. Select a product of your own choosing. Now produce a number of sketches that morph the product into different forms that meet the following specification requirements: urban, military, sophisticated, eco, value, and slow. Show each sketch to colleagues to see if they can guess the design keyword.

8. "Some products can be so good as to simply go unnoticed." Find a product that you think is beautifully designed but is perhaps so commonplace, or so discrete, as to be largely invisible and unappreciated.

Production

The process of generating product ideas and developing concepts is usually relatively cheap compared to the process of getting the product into production. Money is needed for research and experimentation, machinery, tooling, setup, assembly, storage, and testing to get the manufacturing right. Mistakes and changes can also be expensive and difficult to correct so it is vital that the design is fully optimized as it enters into the production phase. This chapter looks at the prototyping and preparation needed as a prelude to manufacturing, considerations around the choices of manufacturing processes, and the operational issues relative to production that can influence the success of a product.

Pre-production

Preproduction is about making sure that the idea that has been developed into a draft product is fully realized and fully resolved, before it is given the sign-off to enter into the manufacturing stage. Attention to detail is critical in ensuring that working drawings, working models, costs, and performance are all optimized because it is hard and expensive to make changes or rectify mistakes after full production is commenced.

Visualization

Good product visuals help a designer to finalize a concept and to sell the emerging design to colleagues, managers, or investors as a prelude to entering the production phase. Numerous free or professional graphics packages can be called on to produce digital representations of the form and shape of the concept. Raster-based packages and photo editors work with pixels and

Figure 4.1
Copper shade pendant light by Tom Dixon. The light is made from a polycarbonate sphere which is vacuum metal-lized with copper, or copper and brass.

can be rendered to near real images. It is harder to achieve this with illustration-type, vector-based packages which work with curves rather than pixels, but vector-based images are easier to manipulate (scale or morph for example), need lower file sizes, and usually produce cleaner images with standardized color tones for professional printing.

Graphic tablets allow designers to draw forms with digital pens providing a more intuitive and flowing movement than a computer mouse. The ability to save images in appropriate formats is also important. Usual formats for graphics applications include JPEG, TIFF, BMP, or PNP. TIFF (Tag Image File Format) is often the preferred format for professional reports and flyers because of the flexibility and simplicity it offers in compression and instruction.

"Be a yardstick of quality. Some people aren't used to an environment where excellence is expected."

Steve Jobs

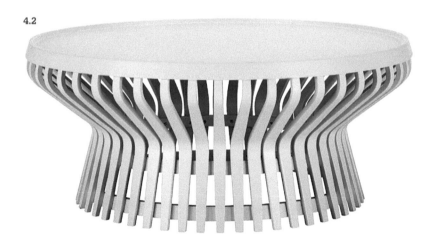

4.2

Figure 4.2
Deena Low table for Habitat. Symmetry is easily generated within digital packages helping to quickly visualize complex three-dimensional forms such as this structural coffee table made from an array of color-lacquered, slatted beech plywood.

Computer aided design

Technical drawings are needed to communicate information such as materials, shapes, and tolerances in a format that allows a product and it's components to be manufactured. This is predominantly done through solid modeling in CAD packages. It is possible to render images in CAD to near photorealistic finishes and even to view CAD models as virtual prototypes in augmented stereo reality.

There is, however, more to CAD than visualization. Modern CAD packages have revolutionized the design process. Many CAD packages allow assembly and interferences between parts to be checked easily, and more complex engineering function such as motion, stress, or fluid flow can be evaluated simultaneously. Packages that use three-dimensional parametric modeling allow features in the design to be linked (referred to as *design intent*). If, for example, it is important to maintain the proportion of height to width in a design, this constraint can be identified so that a change in the dimension of one automatically changes the dimension of the other. It's possible then to change both the length and the width by altering one parameter; wholesale changes can be made if lots of features are linked. This principle of **parametric design** means considering CAD as a feature-based modeler rather than a geometric one. This means being able to make simple changes in solid model CAD drawings without having to start again if a design needs to be changed. It means libraries of standard feature parts can be used, saving design time. It also offers the designer a chance to produce nonstandard designs quickly in order

4.3

Figure 4.3
Levitator floating speakers by James Coleman. A magnetic field in the base unit of this concept design causes superconductors in the speaker to launch and levitate like an alien spaceship, as illustrated in these CAD visuals.

4.4

Figure 4.4
Segway development. Computer aided design (CAD) drawings offer an advantage because they can perform a number of functions, from computing volumes or listing parts to calculating stresses and mold flows. In particular, they can be "unwrapped" into orthographic projections or exploded, as is the case here, to provide essential manufacturing information.
Photograph: segway.com

to meet an individual customer's specific requirements, which, as noted earlier, is an increasingly important market criterion. CAD drawings can still if necessary produce the relevant two-dimensional working drawings in orthographic projections (the top, side and end views) but computer-aided manufacture (CAM) also allows CAD drawings to be linked directly to the manufacturing process and equipment.

The use of CAD requires different perceptive skills to those used in drawing. Freehand drawing requires hand and eye coordination along with an individual style—usually left to right and working backward or forward. A computer-drawn image works in planes and layers and requires less coordination and style but calls on more spatial awareness and levels of abstraction. It's also important to consider saving drawings in the correct format so that CAD data files can be exchanged with clients or manufacturing partners, or even within a single company if CAD is just one part of a series of computer-aided processes within a corporate development process (sometimes referred to as digital product development or DPD). Normal CAD file formats include IGES, STEP, DXF, and DRG.

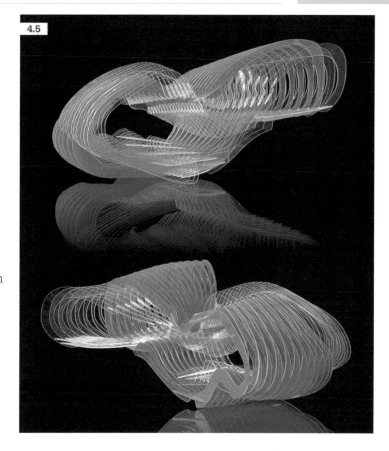

4.5

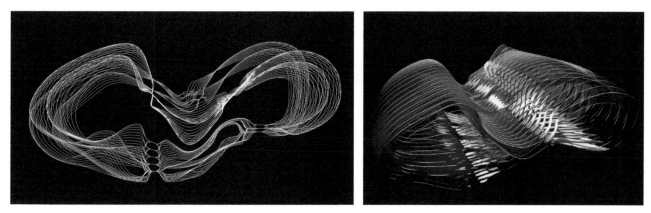

Figure 4.5
Achim Menges Morpho Design Experiment No. 1. This is an example of collaborative design linking artists and technologists. This experiment uses digital means to generate complex geometric planes. The curves of these planes are then explored and linked through further analytical software. The programming can be set to develop increasingly complex, morphing shapes yet be bound by the reality of what can actually be manufactured by laser cutting.
Photographs: Achim Menges

Prototyping

Functioning products with moving components can be modeled in CAD but are often best served by a physical prototype. A **prototype** is an early stage working model that captures the key elements of a design and resembles the intended final product. It can highlight the way a developing product works in a way that an image or static model can't. Issues of movement, tactility, weight, sound, fluidity and function can be simulated and assessed. Even Rapid Prototyping techniques which have made it simple to print a model from a CAD drawing does not replace the benefits that can come from crafting a model from scratch where feel and experience can help to optimize the design.

Rapid prototypes

Rapid prototyping (RP) is the name given to a group of techniques that normally use a CAD-based model to directly generate a physical 3D model. This result is achieved in effect by "slicing" the computer-generated solid model using mathematical methods and reproducing each layer in physical form on RP machinery to build a solid model. Build layers are typically anywhere between 0.025 and 0.15 millimeters thick so there may be some slight ridges, but they are often barely detectable. The machinery can work in a variety of ways including printing paper layers, printing layers of starch or polymer powder like an ink jet printer (3D printing), using lasers to scan and solidify layers of resin (stereolithography), or sintering polymers or even metal powders using lasers (selective layers sintering or SLS). Rapid prototyping is ideal for quickly making products that can be visually inspected before they are developed further and any subsequent redesign can be just as quick and just as instantaneous to review.

Costing

It is vital to have an idea of the development costs of a new product, reviewed against the anticipated market size and financial gain to ensure that it is still viable to proceed into the

Figure 4.6
Napshell working model. The Napshell has been developed by Markus Abele and his team at the University of Stuttgart to aid and legitimize for future working places power napping, a social habit that varies from country to country but that is believed to promote creativity and health.
Photograph: Napshell™, Courtesy of Napshell™

4.7

Figure 4.7
Hidden.mgx vase by Dan Yeffet "Jelly Lab" for Materialise. The Hidden.mgx vase is made from polyamide (nylon) using SLS, a design that could realistically only be made by rapid prototyping. Yeffet defines his work as challenging limits and boundaries, working as an explorer and adventurer.

4.8

Figure 4.8
Mini. Even industry can get costing wrong. For example, the original Mini has iconic styling and status, but greater attention to production and costing details should have ensured a much greater profit margin. Many people think every Mini sold actually lost money.

manufacturing phase. If the expected costs are high and the return is too low, then it is safer and easier to simply invest the money into a savings account. Therefore, it's important to have effective decision-making points in the design process along with accurate data and costing methods.

Most costing methods involve accounting principles of direct costs and indirect costs. Direct costs (also referred to as variable or prime costs) are the costs of a product that vary directly with the number of units sold. This typically includes for example the amount of time that is spent by people making one unit of product (labor cost) and the cost of material for each unit (material cost). If this data isn't accurately known, then a **standard cost** estimate is used.

A product should also include an allowance to cover the expenses of a business that must be paid whether or not it makes anything at all. These include items such rent, salaries, equipment, machinery, heating, and lighting and are referred to as indirect costs (also referred to as fixed costs of overheads). Tools and dies could be categorized as either direct or indirect costs, but if they are used for a specific product, they are usually referred to as a direct cost (or a variable overhead). The cost per unit depends on the cost of the die and the number of parts it can produce before it wears out.

Direct costs are relatively easy to calculate, but accountants can use a number of techniques to calculate indirect costs. It's important to understand this because the method chosen can affect the price and, therefore, success of a product. The methods include

- Fixed overhead absorption rate costing (FOAR)
- Marginal (or contribution) costing
- Activity-based costing (ABC)

Total costs of a product are then worked out using a combination of the direct and indirect costs. Profits start to accrue only after the number of products made has generated enough money to cover the total costs (referred to as the **break even** point). This aspect of design is often missing from student work, which seems to occur partly because it forms a low priority compared to the demands of making a product work well and look good. It is also partly the result of the difficulties that arise in generating indirect costs for a notional business. Students should understand that costing is paramount in industry and that every feature they add to a design, and every line they put on a drawing, adds cost and needs to be justified. Without thinking, it's very easy to turn an injection-molded part from a low-cost, two-part die into an unviable high-cost, multiple-part die with low and expensive cycling times.

"We shall make electric lighting so cheap that only rich people will be able to afford a candle."

Thomas Edison

Tolerances

The key principle behind mass production is that individual parts are not made uniquely to match and fit each other but are instead both interchangeable and replaceable. For example, parts produced at different time of the year can be selected in the knowledge that they will still fit together. Some of the earliest examples of this were the sheave blocks (pulleys) made for the British Navy. Marc Isambard Brunel (father of Isambard Kingdom Brunel) and Sir Samuel Bentham developed a system of simple production operations that allowed over 100,000 blocks to be made per year. The system was not widely copied in the United Kingdom, but it was taken up in the United States and popularized by the automobile production of Henry Ford.

To enable interchangeable fits, component dimensions have to be specified and defined against a set reference point (the **datum**). The degree of accuracy described and the allowances made for variations in the dimensions that will still allow parts to fit together are referred to as the **tolerances**. These too have to be specified. Generally, the tighter and more exact the tolerances are, the more likely the fit will always work, but the more expensive it will be to produce with profound effects on costs and, therefore, product sales. The **process capability** and accuracy of the production machinery also need to be considered. Geometric dimensioning and tolerancing products and components are, therefore, a balancing act between performance and production and involve identifying those features crucial for the performance of the product (**critical-to-function(CTF)** tolerances).

The challenge for the designer is to design parts that can be produced in a way where tolerances are simply not (or less) important. Statistical measurement, right first time, zero defects, **robust design**, and Taguchi method are some of the many supporting principles that are badged under the umbrella term of *quality* and that have been widely taken up by companies across the world.

Value analysis

A good designer should constantly seek to optimize a design throughout the development of a product. This process might be ad hoc, looking through the different elements of each design or skimming from one area to another. It can be done quickly by a designer (if the priority is function rather than costs) or without clarity if there are many issues being considered at the same time.

Value analysis (or value engineering) is a more systematic process of assessing a design to see if improving the product's key functions can increase its value or if removing unnecessary features and functions can reduce the costs. It does this by looking at the design holistically rather than in parts and working with concepts of scoring value and cost. In this way, it can reduce the amount of waste and materials used, lower costs, and improve profit margins and is, therefore, an important tool in creating commercially successful, mass-produced products. It often forms a part of a company's process of continuing improvement and in some cases (e.g., in the United States) it is required by law.

FINANCE

A careful approach to costing and decision making is required. Too much attention to cost at the wrong time can stifle imagination and development. Not enough might result in an overblown development costs and overpriced products. One of the problems is that the amount of money required to finance raw materials, equipment, staff, and marketing for **new product development (NPD)** can be considerable, but the returns can never be guaranteed. Companies can then be unwilling to take the investment risk. Raising finance can then be one of the biggest barriers that designers face in getting their products to market.

Performance testing

Earlier, the term *user* or *product validation* was used to describe the reality check to see if the emerging design concept actually meets the requirements the consumer originally envisioned. *Product verification* is the term used to describe if the product actually meets the design specification. For some products, independent testing and certification is a compulsory requirement. It can save money to think issues through before expensive physical testing and a failure mode and effect analysis (FMEA)—seeking to identify potential failures in design and/or manufacturing including actions required to prevent failures—is a useful matrix for doing this.

From large multinational companies to design students, time and delivery pressures can make performance testing an overlooked stage that can slip to the back end of the project. Product longevity can be particularly overlooked by students. Having worked hard on a design, a designer can also feel that the design is optimized and be reluctant to hear any negative comments. However, consumer comments, whether negative or positive, along with performance and (if necessary) destruction testing, are all a vital part of

the design process. It's better to know the weaknesses now, and the likely **failure rate**, before entering the manufacturing stage where mistakes become very expensive to rectify. Designers must be able to learn from results, listen to criticism, argue a case if appropriate, and accept a change if it's necessary.

Checking the safety of a product is an essential part of the product-testing regime. Risk assessment is a systematic approach to exploring the hazards that people might be subject to in using (or connecting in any way) with a product. These should include readily recognizable hazards such as sharp edges or high temperatures as well as less obvious hazards such as entanglement, entrapment, vibration, and noise. A risk assessment combines this list of hazards with an evaluation of the likelihood and outcomes if any of the hazards actually happen. The aims are to reduce the hazard or ideally eradicate it completely. In undertaking a risk assessment, it is important to recognize that products may be used by people who do not constitute the original target user profile or by people who are stressed, tired, or distracted. Products may also not be used in the manner for which they were intended.

BEING MORE COMPETITIVE

Kaizen comes from the Japanese word for improvement and its use today is concerned with the continuous improvement of a company's operation. Kaizen involves a range of issues, from waste reduction to standardization, and its method include individual responsibility, trust and communication. This is a different philosophical approach to more traditional authoritative business control.

Environmental benchmarking

Even though most people recognize the adverse effects that products and consumerism have on the environment, designers can seem to struggle to know how to react to the challenge to make products more sustainable or eco-friendly.

Environmental benchmarking is a tool that can help designers respond positively to the environmental agenda. Carbon footprints, eco webs, eco rucksacks, and MET (materials, energy, toxicity) matrices are some of the many emerging environmental benchmarking techniques available. These

Figure 4.9
Quin.mgx lamp by Bathsheba Grossman. This sculpted dodecahedron-based lamp is made from poly amide using SLS rapid pro-totyping. Light interacts with the swooping curved surfaces to pro-duce areas of light fusing with areas of shade and bright spots. This design can easily be tested and refined using RP.

4.9

can allow designers to understand and compare the environmental consequences of different design decisions and can point to the environmentally impactful areas and outcomes of a product that a designer can improve on. A **life cycle assessment (LCA)** is the most rigorous method because procedures are standardized through ISO 14040 series. It also looks at the full range of environmental impacts throughout the lifetime of a product and considers how any design changes might have an indirect effect on other parts of the life cycle.

4.10

Figure 4.10
Bathtub couch by Max McMurdo for Reestore. Recycling generally refers to reusing the same material for the same purpose, such as using a milk bottle again to sell another pint of milk. It can also mean using the material again so that it can be used for other purposes. In this example, a cast-iron bath has been reused and turned into a sofa.

Case study
Assa Ashuach

Assa Ashuach was born in Israel in 1969, and since childhood, he has loved to make seemingly impossible things from bits of scrap. After studying industrial design at the Bezalel Academy in Jerusalem and at London's Royal College of Art, his attention moved from bizarre scrap to more high-tech forms of manufacture. For Ashuach, design is not just about form, function, or emotion, it is about challenging and modifying traditions based on modern technologies. Now based in London, his new furniture and lights were featured at the London Design Museum, the London Design Festival, and Frieze Art Fair. He has received numerous awards including those from the Design Museum, the Esmée Fairbairn Foundation, and red dot.

One characteristic of his work is collaborations with groups such as Materialise and ComplexMatters in the development of a new approach to materials in design, developing materials that are at the front end of the design process rather than slotted in at the end. Another characteristic has been his use of computer programs rather than straightforward CAD models to instruct rapid prototyping machines. The computer programs are a form of artificial intelligence, programmed to maximize the aesthetic and optimize the strength-to-weight ratio. It is a process that he likens to a growing process enabled by a DNA code. Ashuach refers to this as digital forming® and has created a company that develops products based on these principles. It's not just that the thinking behind his ideas is clever, but the products that

emerge are usually exquisite and seductive. More recently, he has founded visionary companies such as Digital Forming Ltd and UCUDO Ltd that enable individual consumers to engage in their own way with his digital forming ideas.

An interview with Assa Ashuach

How did you first become aware of and interested in design?

I was born and grew up in a kibbutz in Israel on the Mediterranean coast. Growing up in a kibbutz in the early 1970s was the best environment for a child to create and play with the exciting old machinery and exotic scraps. Visiting the carpenter, metal workers, car mechanics, and shoemaker, I would always end up with great collections of bits and pieces. I used to

Figure 4.11

Omi.mgx lamp. The Omi.mgx light is one of the first products to be made immediately available to consumers as the output of an SLS process. The lamp is made as a single nylon part with flexibility that allows it to be twisted and bent like an amorphic mutation, creating different shapes and moods. This behavior offers versatility in that it can be personalized and manipulated easily to create different sculptural sensations, space, and moods. Later work, such as the AI Light, include sensors that enable the light to morph itself as it hangs in space, reacting to purposeful or natural changes in light, sound, or movement. 3D printed objects that change shape over time through design or intelligent materials are often referred to as 4D printing.

ask my friends to tell me their greatest material fantasies, so I could have the pleasure of cracking and inventing another impossible thing.

How important is experimentation with technology to your work?

Complex technology helps to simplify design. Alias Studio tools (3D software) gave us the right tools to restrain and simplify a 5-meter-long line. SLS technology helps us to avoid elements that are unnecessary for the design but essential when production is done with plastic injection. The Upica sofa is a composition of four lines that creates one compound surface. It is a very slim and large surface that supports itself. If a few millimeters of the surface-tight shape were changed, the sofa would collapse like paper.

How did the Omi light project and your relationship with Materialise come about?

In 1999, I designed the first Omi lamp made from 120 disposable foam plates. The 120 plates were attached with a rubber band, which allowed the whole bunch to move and to create a type of biological worm mechanism. I showed this lamp at Salone Satellite in Milan 2000. It got very good reactions, but it was very difficult to produce in conventional ways. The first meeting with Naomi Kaempfer, art director of Materialise, was just after my RCA graduation. My experience with 3D and RP technologies together with Materialise.mgx, a young experimental brand, was the perfect match. The first project we worked on was Omi.mgx. It was to take the impossible design into production in the only possible way.

4.12

Figure 4.12
Femur stool. The femur stool is another example of four-dimensional printing. It is designed to maximize its weight-bearing capacity using the minimal quantity of materials and has the ability to conform itself to the weight and posture of the person who sits on it.

Manufacturing

For most designers, manufacturing is about selecting the right processes to enable large-scale production of their ideas. For some designers though, manufacturing is also about experimenting, perfecting, or crafting these processes as a route to new product innovations.

This section looks at manufacturing processes with suggestions as to how they influence design thinking. There are hundreds, if not thousands, of different manufacturing techniques. To help understand them, however, it's possible to categorize them into one of four principle mechanisms: (a) extracting material from solid lumps to form the required shape (machining), (b) sticking bits together to get the desired shape (fabricating), (c) pouring nonsolid materials into a shape (molding), or (d) forcing solid materials into a shape (forming).

**Figure 4.13
Spun chair
by Thomas
Heatherwick.**
Traditional
metal spinning
techniques have
been applied to
the design and
production of the
spun chair.

4.13

Machining

Chipping flakes from a solid stone mass is a production technique that dates back to the creation of man's earliest products, such as axes and knives. Today's material removal (or reductive manufacture) is more likely to be achieved via lathes, milling machines, drills, and grinders. Spark erosion (electrodischarge machining or EDM) and electrochemical machining (ECM) are among a number of other less obvious material removing techniques that offer interesting design options. The principles of ECM have, for example, facilitated the revolution in the manufacture of microchips.

These processes are flexible and accurate and can produce complex forms, particularly when used in conjunction with computer numerical control (CNC). CNC can be programmed directly from a design via computer-aided manufacturing or from parametric programming. They can be set up so that they are self-monitoring with autofeed and autotool changes, allowing the process to run overnight in lights-out factories and without operators. For the purposes of mass production, however, machining tends to be considered a slow option and the offcuts and swarf produced also make it a wasteful and, therefore, expensive process.

4.14

Figure 4.14
Wire chair by Tom Dixon. Inspired by the simple paperclip and designed to meet the needs of a specific market sector, this welded stainless steel wire chair can be used indoors or outdoors and forms part of a furniture series designed by Dixon that is based around the concept of lightweight but functional wire.
Photograph: Tom Dixon

"Furniture manufacturing in plastics requires very costly machinery, which the Danish market is not big enough to justify. Or so they say. But show me a plastics manufacturer who dares to take on the experiment."
Arne Jacobsen

Fabricating

Fabrication describes production via the assembly of different parts. A designer electing this form of manufacture has at his or her disposal an array of processes such as clip-fitting, screwing, welding, and gluing. Each process has its advantages and disadvantages. Gluing, for example, can be quickly applied and provides waterproof seals and invisible joints that have an even spread of load around the area of the seal. Glues have also benefited from advances in material science, and the range of traditional natural and synthetic adhesives has been expanded with the development of drying, contact, thermoplastic, reactive, pressure-sensitive, and light-curing adhesives. All of this gives the designer a much wider range of options, and products such as bicycle frames and even airplanes, which might have been welded in the past, can now be glued.

Gluing, though, can be toxic, subject to environmental degradation and can make recycling more difficult; these factors may prompt designers to look towards the other fabrication processes. Process like clip-fitting or screwing are generally more environmentally friendly because they allow products to be easily taken apart and reused or recycled at the end of their working life.

Molding

Molding processes shape a pliable material into the form of a mold or die. These processes include sand-casting, die-casting, investment-casting, sintering (powder metallurgy), injection molding, compression molding, blow molding, rotational molding, and vacuum-forming. In each case, the material is usually

4.15

Figure 4.15
Little Tikes Cozy Coupe©Car. The Little Tikes Cozy Coupe©Car has been a children's favorite for over 30 years, which is remarkable given the way society, technology, and fashions have changed over these years. It is rotationally molded with double walls for strength and impact resistance. Toy design has often led the way in exploring simple, safe, and quick assembly techniques.

in liquid, pellet, or powder form and is cast by gravity or under pressure into metal dies or into wax-, ceramic-, or sand-formed cavities. Each process varies widely in the size, quality, cost, accuracy, and material that can be cast so the process needs to be carefully matched against the design criteria.

Injection molding of polymers is a popular manufacturing process among designers because it is relatively cheap and can make small or large items with reasonable accuracy. Injection molding can also enable operations such as insert molding (like embedding a metal screwdriver blade into a polymer handle), directly applied graphics through in mold foils, or multi-shot injections to build up layers of different polymers to give unique feel and looks.

Reaction injection molding (RIM) is a similar process except that it uses a combination of component materials that have a chemical reaction within the mold itself. A typical result might be a rigid shell material with a rigid foam-type internal structure, which gives a product such as a car bumper good lightness and strength.

Forming

Forming is the shaping of a hot or cold solid material. It includes traditional blacksmithing and forge work as well as more industrialized processes such as rolling, extruding, high-energy-rate forming, pressing, stamping, and drop-forging. Forming offers distinct advantages to the designer. Forces of 160 MN can be applied in the transformation process, and accurate parts can be produced in thinner sections than can be achieved by casting. Whereas molecules in molding processes are arranged in random patterns, in forming there is also scope for aligning the molecular grain into specific directions allowing the designer the ability to build strength in key areas.

Where large numbers of units need to be made, forming is an easily automated, high-speed process involving little preparation and waste, making it ideal for mass production. Nails, crankshafts, car body panels, hinges, fire extinguishers, drink cans, and cutlery are among the range of products that are made by forming processes.

4.16

Figure 4.16
Zanzibar Tower ZTR (left) and Hot Spring (right) radiators by Bisque. The Bisque radiators Hot Spring (by Paul Priestman) and Zanzibar (by Talin Dori) push the boundaries of manufacturing in steel, using round and square chrome tubes to create innovative radiators.

Incorporating new techniques

Advances are frequently being made in manufacturing techniques and technology, and these can trigger innovative new product concepts. Examples from the past include the iconic seating created by the pioneering Eames brothers, who used the bent wood techniques developed by Alvar Aalto through to the pioneering stacking chair developed by Robin Day using the then new process of injection molding. Nowadays, designers might look at new techniques that include growing metal forms from a single crystal, using biomaterials to grow plastic parts, or spinning nanoparticles like a spider's web.

Even though new developments offer opportunities, some environmentalists and designers suggest that traditional craft manufacturing techniques offer a more satisfying activity and a more sustainable future. This argument between traditional and modern manufacture has been seen throughout time, for example, from the violent anti-technology Luddite movement to the traditional values of the arts and crafts movement.

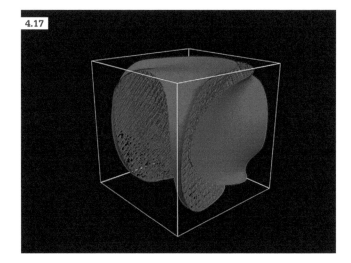

Figure 4.17
AI Stool.mgx for Materlise.mgx by Assa Ashuach. The AI Stool was "grown" in free space emulating the biological structure and mechanism of human bone from a DNA code using programming techniques developed by Ashuach. The code optimizes both for strength and aesthetic. The usual process of product optimization might be to calculate stresses and strains first and to refine and shape the design afterward, but the Osteon chair works this process in reverse. Computer code was used to determine the best internal structure needed to support the chair's required and finalized surface form; this structure was then printed out by laser sintering. It is estimated to use one-third of the material that might conventionally have been used.

Rapid manufacturing

Additive manufacture is an umbrella term for the tiny layering fabrication processes used in rapid prototyping. It is referred to as *rapid manufacture* when the RP process is used to directly produce manufactured parts (or molds) rather than prototypes.

The process is capable of making parts that can be difficult to make in other ways and offers designers opportunities to design new products that can realistically be manufactured only through rapid prototyping. This includes parts without radii or taper, parts within parts, cavities without access, interlinked shapes, metal parts that phase change or are not homogenous, or those with wispy fill sections. The process is finding a niche role in aerospace and medical production and in small batch production.

The reality is that rapid manufacture is far from rapid compared to many traditional manufacturing techniques. As the size and speed of the technology increases however, and the costs reduce, many think that it will become one of the principle manufacturing techniques of the future.

Figure 4.18
Freedom of Creation 610 Lamp by Jiri Evenhuis and Janne Kyttänen. These lamp designs feature advanced materials such as Keronite and laser-sintered polyamide, combined with inspirational thinking based on natural mathematical series (such as Fibonacci). The company's products are stored digitally and can be downloaded for viewing in virtual reality environments or for use in rapid manufacturing techniques.

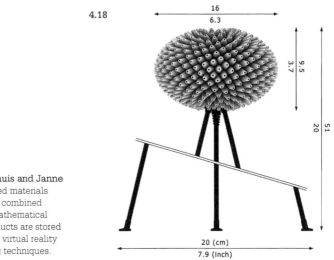

4.18

16
6.3

9.5
3.7

51
20

20 (cm)
7.9 (Inch)

Case study
Tom Dixon

Tom Dixon describes himself as "a self-educated maverick whose only qualification is a one-day course in plastic bumper repair." His early life saw him experience a variety of different cultures, and his early creative works are marked by his interest and hobby in welding, which allowed him to create structures quickly from scrap (one of which led to his collaboration with Italian furniture company Cappellini and the realization of the celebrated S chair). Dixon is regarded as a highly creative designer and has many product awards to his name, but he prefers to be considered an industrialist. He has been scathing toward British manufacturing and design bodies for not supporting British design and innovation sufficiently.

Dixon's experience with both Cappellini and his own Eurolounge enterprise represented a journey away from the realm of jobbing industrialist creator to that of more business savvy designer. This direction went a step further when he was taken on as head of design in the United Kingdom at Habitat. Placing a "maverick" into a mainstream, corporate, retail-driven business was risky for both Habitat and Dixon. Dixon, however, was pleased to learn more about mass production manufacturing and costing and to have access to a worldwide basket of design possibilities. In fact, he has described his decision to learn more about developing commercial products as a state of "growing up." In return he gave Habitat a makeover that was uniformly considered highly successful. Among his initiatives were

the reintroduction of design classics and the fostering of new design talent. Dixon later became Habitat's creative director and was later awarded an OBE for his services to design.

Dixon now works largely as an independent designer and has not compromised his individuality in order to be commercially successful. He continues to be innovative, creative, and controversial, yet he can design successfully for both high-end and mainstream markets. He designs lighting and furniture under his Tom Dixon brand and collaborates on a range of design projects through his Design Research initiatives. He has a continuing thirst for knowledge, too; ask about his current work and he talks not just about ideas but about his interest in engineering, marketing, manufacturing processes, and digital production.

"I am still mainly motivated by materials and processes but these preoccupations evolve. I am currently exploring blow molding, vacuum metalising and computer controlled manufacturing systems."

Tom Dixon

4.19

Figure 4.19
Slab chair. Deep brushing exposes the grain of this oak chair, which is enriched with black lacquer.

4.20

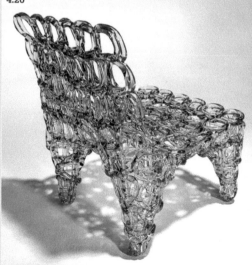

Figure 4.20
Fresh Fat Easy Chair. A single strand of hot, extruded plastic is hand-formed into loops that solidify when cooled. The uniqueness of each chair challenges the more normal concept of polymers as a mass-produced, throwaway material.
Photographs: Tom Dixon

Operations

One single manufacturing machine on its own is unlikely to produce the vast range and quantities of consumer goods; instead, a host of equipment, people, and suppliers is usually required. The organization of this "materiel" can significantly affect the costs of making a product and ultimately, therefore, its success. This organization is referred to as *production operations* and is a vast subject in its own right. This section illustrates, however, some of the key principles and the impact upon the work of designers.

Design for Manufacture and Assembly (DFMA)

Selecting an appropriate manufacturing technique is a key task, and designers must ensure that it and the design are well matched. If selecting a machining process, for example, designers might choose a softer material that is easier to machine, or they may provide a flat rather than a sloping surface to make it easier to drill holes. Or if molding, they might provide sloping sides and rounded edges to help extract a molten part from a mold after it has cooled. These design features are vital in getting a product manufactured at the lowest possible cost. They should be made at an early stage in the design process rather than during the manufacturing stage when it is discovered that something cannot be made easily, or even at all! This consideration process is referred to as design for manufacture (DFM).

The process of changing a product so that it can be put together more easily is referred to as design for assembly (DFA). It is every bit as important as DFM since much of the production cost might actually be accrued at the time-intensive assembly stage. Considerations might include providing lifting lugs for heavy parts, holding points to keep it still, symmetry to make parts easier to pick up,

Figure 4.21
Door handle by Dieter Rams. Matte, brushed, nickel-plated and clear-lacquered door furniture. Simple lines, radii, and curves combine elegant styling with ease of manufacturing in this design.

4.21

or a standard range of bolt sizes rather than a mix of shapes and sizes.

Design for disassembly (DFD) has become important in the role of recycling or repairing products. In some countries, it is a legal requirement to consider what is done to a product at the end of its working life. For example, The Waste Electrical and Electronic Equipment (WEEE) is a largely global directive specifically aimed at improving recycling rates. It means that manufacturers need to consider what happens to electrical and electronic equipment at the end of their working lives.

Modular design

The principle behind **modular design** is to view a product or a range of products as a system that can be divided into smaller parts. These smaller parts are then designed to a common format (or standardized). For example, a company producing three products A, B, and C, with three different motors, might find it more cost-effective to use the same motor for each different design, even if this means using a larger motor than the product needed. This is because the one motor can be produced in larger numbers using the cheaper techniques of mass production (or obtaining larger supplier discounts).

This principle is often obscured if designers are working in a large organization and on individual products, particularly with a long list of components and parts. The issue is how to capture the systems and commonality across the range of products, and this requires the designer to be able to access good data and knowledge management systems. Modularity in design is also a key ingredient in allowing consumers to choose their own combination of features and, therefore, have more choice in the individual nature of their particular product.

4.22

Figure 4.22
JCB teletruck. The teletruck has combined the strength and ruggedness of a digger with the flexibility of a forklift to create a hybrid vehicle that incorporates a telescopic boom instead of the standard lifting platform. Innovative ideas still need courage to develop, but development costs can be reduced by drawing on existing products and components.

Manufacturing conformance

The levels of perfection required for products on the market today compared to previous years are very high, and imperfections are not usually tolerated. Manufacturing conformance refers to the testing process that ensures components are manufactured to the standards and specifications that have been predetermined. This can involve measuring variables (features with quantified values such as length or weight) or more subjective attributes (such as color) using generic or special-purpose tools and equipment.

Decisions have to be made as to when, where, how often, and how many parts need to be assessed. **Six Sigma** is a term that refers to the fact that the majority of outcomes in a nonbiased and random normal distribution fall within plus or minus 3 standard deviations from the average outcome. Six Sigma in business refers to an approach and a set of techniques used by some companies including Motorola and General Electric for getting 99.99966 percent of the features of a component part free from defects. Other companies, however, have argued that Six Sigma stifles creativity and research. Conformance is, therefore, another juggling act of trying not to slow production down and encumber it with paperwork and tests balanced against the cost of scrap and reworking if parts are made incorrectly.

Supply chains

It is relatively easy now to e-mail a sketch from one country to another and to have a product made and shipped in just a few days. Behind this simple façade, however, lies a complex network of financial systems, communications technologies, transport operations, and trade agreements. The globalization of designing, sourcing, manufacturing, transporting, and trading across the world involves multiple links and networks referred to as supply chains, and the management of these is complex and sophisticated. Even if designers have elected not to enter the global marketplace, whether they should use the global supply chain to get the right materials, manufacturing, and know-how at the lowest costs needed to deliver their ideal product—knowing that these chains might be contributing to environmental and sustainable damage—is another issue that designers must consider.

Just in time

On the face of it, the process of reducing stocks of raw materials and components in a factory and choosing instead to have these items delivered as and when needed (usually just in time or JIT) is a production technique. It does, though, have a profound relationship with

Figure 4.23 Quinny Buzz pushchair. The Quinny Buzz looks striking, and its design has met many user needs including a seat that can be removed and used in the house or the car. The seat can also be used reversibly in the buggy itself so that the child can face forward or backward. The buggy's unique triangular frame makes it lightweight, maneuverable, and capable of unfolding automatically.

4.23

design. As the supply chains become sharper and more efficient at delivering, the production batch sizes can become smaller and smaller. If production is flexible enough (and the design is modular and robust), then the batch size can eventually approach a unit size of one. In other words, products can be mass-produced but on a one-by-one basis. If this is linked to sales, then products are effectively made to order for each consumer with no storage requirements and no waste. The savings that result from reduced waste and uneven workload demands are referred to as **lean manufacture**.

Globalization

Over the last forty years, world exports have doubled as the possibility of working, operating, and trading beyond the immediate national boundaries grows, spurred on by technical and political developments and the need to compete on a larger scale. This process has included outsourcing manufacturing to countries with lower cost bases for cheaper production. This brings up another debate around globalization. Does it lead to the spreading of wealth, or does it lead to the decline of local cultures and a decline of local authority or derogation of employment rights, as well as contributing to global environmental damage? Designers should take an opinion on this question. Should they aim to design for a global market—and risk the sustainable future of the planet—or should they stay localized at the possible expense of their own livelihood?

4.24

Figure 4.24
POÄNG chair by Noburu Nakamura for Ikea. The POÄNG chair is made from bent and glued beech wood veneers that are stained with clear acrylic lacquer, solid beech rails, polypropylene-support fabric, and leather seating. It is one of Ikea's top-selling lines and has been purchased by many of the 583 million visitors to the company's stores worldwide.

"Globalization has changed us into a company that searches the world, not just to sell or to source, but to find intellectual capital—the world's best talents and greatest ideas."

Jack Welch, formerly chairman and chief executive, General Electric

Case study
Salter

Salter Housewares has been in the weighing business since the 1700s and is a leading player in kitchen scales with around a 40 percent share of the U.K. market alone. History can, however, count for little in the commercial world where a failure to innovate can spell disaster. The kitchenware market is a prime example with new products being released on a regular basis, not only by traditional competitors but also by new, dynamic companies that are able to enter the market because of the relatively straightforward technology involved.

The primary function of scales is, of course, to weigh matter, but the designers at Salter also have to meet several other market needs: the need for compactness in busy kitchens, for style in modern lives, for cleanliness and hygiene, and for ease of use. The 1007 model is designed to meet these needs in a number of ways. It has a very individual ring shape that gives it a distinctive and aesthetic form, and the flat, doughnut shape allows the product to be easily stored. The central, circular platform is made of glass, which is hygienic, easily cleaned, and resistant to staining. It is also very functional, allowing ordinary dishes to be used as the measuring bowl or for ingredients to be placed straight onto the platform. The digital readout has unusually large bright digits that are shown on a black background, which gives a very clear display. Finally, the scales also have an add-and-weigh feature that allows ingredients to be added one after the other while continually weighing. This feature saves the need to empty or rezero each time and supports the way in which cooking through a recipe is actually done.

An interview with Karen Brown of Salter Housewares

Are you able to design a single global product?

It makes economic sense to try and design one product that appeals to as wide a customer base as possible, but we are aware of the fact that different markets have different peculiarities . . . . The dial of U.K. scales [for example], will typically include older imperial units as well as metric, whilst a European version will show metric readings only . . . . A modular approach helps to meet those differences.

How do you make sure everything conforms to the original design intent?

We use consumer focus groups regularly to ensure that customer needs are kept to the forefront of the design. We also make a point of seeking feedback from our retailers too, as this can be extremely informative. With regard to actual manufacturing conformance, like most companies we have rigorous test procedures and work to ISO 9001 (Quality Management Systems) . . . . Good communications are a key to ensuring conformity, particularly in ensuring that we and our suppliers are working to the same specifications and requirements.

How quickly does it all work?

Our new product development process typically takes around six to nine months from concept to manufacture. We can reduce this time on occasion if necessary, but this is the sensible speed . . . . The whole development process requires rigid scheduling and attention to detail, but it's not a purely mechanical process. We have a good mix of both young and experienced designers in the team to ensure that we maintain creativity and innovation at the same time as producing practical and manufacturable designs.

4.25

Figure 4.25
Kitchen scales 1007 model by Salter Housewares

Chapter summary

This chapter has looked at the process and concepts behind preparing and manufacturing a product prior to sales launch. Some of the concepts raised such as tolerancing, costing, production, mass production processes, value analysis, and material science can often seem less glamorous for designers who might prefer to leave these details to last or to just leave them completely to production engineers. This is an attitude that can be exacerbated by the ability to manufacture overseas ,which can seem like a low-cost production panacea.

Production is, however, fundamental to design. Even when made in low-cost economies, a product that uses slower methods of production, uses more materials, or takes longer to assemble than its rivals is already at a major competitive disadvantage no matter how well it has been designed. Attention to these details can make or break a product; therefore, designers need to understand and work with manufacturing technologies and systems to ensure their products are easy to make.

At the same time, designers can learn from and push the boundaries of manufacturing as a route to innovating new products. Some of the most exciting new designs seen in shops and exhibitions are also the result of successful experiments and thinking by talented designers using new materials and manufacturing techniques. These are rich areas of exploration for designers, and many designers and design movements have made their names by being associated with these pioneering works.

Assignments

1. Research and identify one new and developing manufacturing technique, and apply it to the design of a new concept product.

2. Select one product and attempt to redesign it to make it 20 percent easier to manufacture without losing any of the product's functionality or appeal.

3. The quality approach to design has in the past been slow and has not always been successfully introduced to Westernized industrial corporations. Why is that?

4. A furniture company wishes to capitalize on the environmental market by creating a new outdoor range to encourage children to be outside more than indoors watching television. They have asked you to design, using this new concept, an outdoor piece that would allow three children to sit and chat together. The design is to be made from natural materials of concrete with hessian hammocks or seating with a PVC cover available for inclement weather. Design the product for them, and then produce a report that appraises the product environmental credentials using a benchmarking tool such as a MET (materials, energy, toxicity) matrix or ecoweb.

5. In groups, debate whether designers, consumers, politicians, or corporations should take the lead role in protecting the environment.

Product launch

Having worked hard to research and understand a problem, to generate contemporary and creative solutions, to develop a concept into a cost-effective and viable product, and to manufacture it efficiently, it can be frustrating—to say the least—to see a well-designed product fail because of poor attention to the process of launching and selling the product.

This chapter looks at the issues of rolling a product out and the selling function that help to ensure the success of the design. In a fiercely competitive world where products can fail more often than they succeed, it is vital to attend to these. The chapter explores how and why the designer has a role to play in these functions even after the product has left the factory gate.

Roll out

This section introduces some of the important issues for designers, as manufactured products move from the factory and into the commercial market place including some of the penalties for releasing defective or dangerous products and some of the ways of protecting your products from being copied by competitors.

Testing

Even though the design should have been validated, verified, and conformance approved so that it is effectively ready to go, there is still a need for testing and final sign off. Does it function as expected? Does it meet its statutory requirements? How does it now compare and benchmark against new products released by rivals during the development process? Has the market need actually changed during the development process (a problem where there have been long development times). Products should be evaluated against a carefully constructed test plan, tested, for example, via in-house analysis (**alpha testing**) or consumer pilot groups (beta testing). Product liability and risk need to be included in the test plan.

Figure 5.1
Dyson DC54 cleaner

5.1

These can be moments of high stress and panic for the design team, and sometimes decisions have to be made whether to launch with flaws or spend more time and money to get the product perfected.

Liability

Most product defects are the result of failures in manufacturing. Inherent design faults account for around one third of product failures, although a significant number of failures are simply due to the failure to provide adequate information for consumers on how to operate the product.

There is often variety in the way that countries approach product liability for products that don't work. Some will rely on the market's ability to self-regulate (i.e., people will not buy products that are faulty), but most draw more heavily on the law. There are normally contractual arrangements between suppliers and consumers that deal with faulty products. For example, EU Directive 1999/44/EC outlines some rules for nonconforming products, which in the United Kingdom work in conjunction with the Consumer Rights Act.

Laws around products that are dangerous can be more onerous and include criminal convictions. Since products may be exported anywhere in the world, designers should work to the appropriate legislation world-wide and not just that of their own country. Actions to reduce liability and risks include making appropriate design calculations, experiments, and tests and by undertaking a systematic risk analysis of the product and its production.

The European Economic Area (EEA) requires some products to be CE or conformance marked to show that they have

WHAT IS A DANGEROUS PRODUCT?

Most countries have a form of legislation that prohibits the sale of hazardous goods. A chain saw might typically operate with around 250 unguarded teeth bites per second. Even though there are undoubtedly dangers in using a chain saw, and many people do get injured from using them, it does not automatically fall foul of usual safe product regulations for reasons such as the explicitness of the danger, through careful attention to safety features and the availability of supporting protective equipment and training. There are, however, complex issues to consider around the purpose of the product, the context of its use and whether the design could be made safer, and even the wording used in the operation manual or in the selling of a product can affect the interpretation of what is safe.

been produced according to the relevant directives and legislation. The US. Federal Communications Commission (FCC) label is required to show that some electrical and electronic equipment is certified to perform within set limits for electromagnetic interference. Parts of Asia and Africa use the FCC label; other countries have their own equivalent such as China's CCC mark or Japans VCCI.

Patents

Developing a new product can be a long and strenuous process, and it can be devastating to see others reaping the rewards from the fruits of your labor by copying your work for a fraction of the effort. The legislative protection given to protecting the knowledge and outcomes afforded by design work are collectively called the intellectual property rights (IPR). Intellectual property rights include a number of forms of protection but those most affecting product design include patents, design rights, copyrights, trademarks, database rights, and moral rights.

A patent, for example, is a form of IPR that aims to protect technical or functional innovation that might include the way a product works, the way it is made or used, or any new materials or manufacturing techniques. The innovations must have commercial value and must be innovative, meaning there should be some originality that is previously unknown (and this means keeping all work confidential).

Some countries, notably the United States and Japan, award patents to the designer who is "first to invent," whereas others reward those who are "first to file" for a patent application. For example, some people consider that Elisha Gray—among others—had the right to be acknowledged as the inventor of the telephone, but it was Alexander Graham Bell who secured the patent first and to whom the invention is now generally credited. The key implication for designers is to not to delay in filing for a patent and to keep progress records in authenticated logbooks.

Figure 5.2
EvoShave by EvoShave Ltd. The EvoShave handle was designed to act as a bridge to connect the fingers directly to the cartridge in order to improve the biomechanics of shaving. This new interpretation of a razor handle function underpins the EvoShave's patent. The handle form allows the small muscle groups in the fingers to flex intuitively over the contours of the skin taking away responsibility of movement from the larger muscle groups of the arm that are comparatively harder to control over small distances. The handle is a two-shot plastic/rubber injection mold wherein the plastic is an ABS polycarbonate mix that is harder than the pure ABS of the cartridge for reasons pertaining to the cartridge clip mechanism. The EvoShave system also comprises a radical cartridge magazine, tray, and reusable carry case design.

5.2

If granted, the patent affords protection against reproduction by others for a period of normally up to twenty years. U.S. companies that lead in the filing of patents include Microsoft, Procter & Gamble, Nike, Goodyear, Black & Decker, Wolverine World Wide, Kohler Company, Apple, 3M, and Ford. In addition to seeking protection, it's equally important to ensure that any of your design work does not infringe the rights of others by undertaking the necessary 'due diligence' investigations of previous intellectual property.

Design Rights

Most countries have an intellectual property right that allows designers to protect the form and shape of their work. These are referred to with terms such as *industrial design*, *registered design*, or more recently in the EU Community *design rights*. Sometimes these rights are granted automatically, and sometimes they must be applied for (registered). More protection is usually afforded to those who register a design, which normally then protects any aesthetic form, including 2D patterns, against any form of reproduction (including accidental copying).

5.3

Figure 5.3
Around Clock by Anthony Dickens
for Lexon

Commercial protection

Rules for intellectual property rights vary from country to country, and this disparity has made it difficult to set up worldwide forms of protection. Securing them can subsequently be a slow and expensive process—very expensive if your applications have to be defended in court. Smaller companies and designers, or those in fast-moving industries, may prefer instead to use commercial methods to protect their products. This includes concealing technical know-how, ensuring sales through brand loyalty, or introducing new and better products at a rate faster than competitors can respond.

Figure 5.4
Dyson technology.
Dyson has a reputation for using and enforcing IPR to protect its design work, but fast-paced innovation and brand loyalty play a strong role in securing their marketplace.

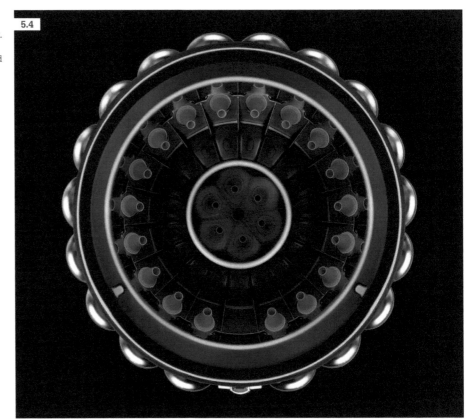

5.4

Knowledge management

5.5

The lesson learned from the development, the design process, and the testing of products should be used to help with the development of the next generation of new products. **Knowledge management** is the term that describes the systems and technologies that capture, organize, and use effectively this information and so forms a vital part of the design function for products that can be complex with numerous parts and components. Even the way parts are named and numbered (**part classification**) and itemized (in a **bill of materials**) becomes of critical importance particularly in large companies where employees and products can number hundreds or even thousands.

Any big company with a range and depth of expertise applied across a range of sectors must have good knowledge management systems in place to avoid chaos, repeated work, or lost opportunities.

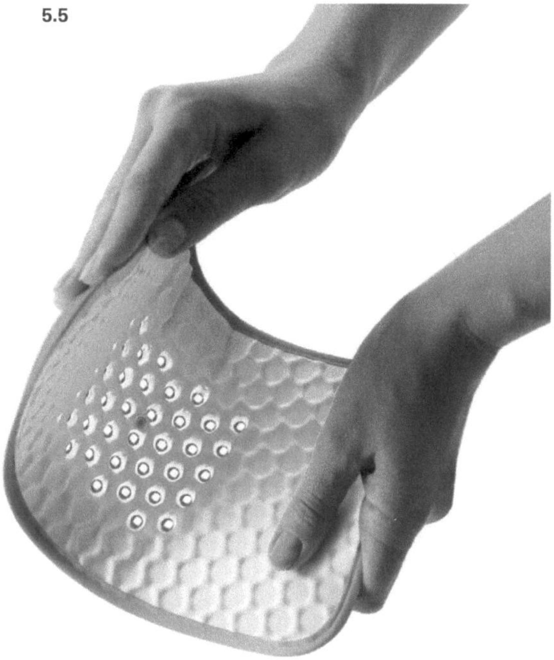

Figure 5.5
Blue Touch by Royal Philips N.V.
The Blue Touch is a healthcare product that crosses over Philips expertise in two main areas; lighting and healthcare. It is designed in response to the idea that blue LED light can stimulate the body's natural pain relief processes with additional nitric oxide in the skin, thus avoiding the need for more damaging UV radiation or chemical treatment.

"No manufacturer, from General Motors to the Little Lulu Novelty Company, would think of putting a product on the market without benefit of a designer."

Raymond Loewy

Case study
Sir James Dyson

Sir James Dyson was born in Norfolk, England, in 1947 and began his design career at what is now Central St Martins College of Art and Design and the Royal College of Art. His product designs include the Seatruck (a high-speed watercraft) and the Ballbarrow, but he is probably best known for inventing the dual cyclone bagless vacuum cleaner, which works on the principle of cyclonic separation.

Dyson is keenly concerned with function in design, resigning at one point from the board of London's Design Museum citing its over-reverence to "style." His determination to get things right is also well documented through his long struggle over five years to get the concept of the dual cyclone vacuum cleaner perfected. This period saw him create over 5,000 prototypes and risk his property and future financial security. What has helped him to take this risk was an inherent belief in the product and the security afforded in the form of intellectual property rights.

The patent for the invention allowed him to take the finished prototype to a range of existing vacuum cleaner companies, although none took up the idea. Eventually, he borrowed £600,000 (US $960,000) to set up his own company Dyson Appliances. Despite being one of the most expensive cleaners on the market, and with very little in the way of sales promotion (relying instead on word of mouth), the Dyson cleaner was a huge sales success.

Patent protection then proved invaluable when designs similar to his began to appear on the market. For example, a new cleaner from Hoover used centrifugal force and air filters to suck dirt from carpets and furnishings. Hoover avoided using the words *dual* or *cyclone* instead referring to *triple vortex* and *amplified spin-cleaning* technology. Hoover denied any patent infringement claiming its cleaner was based on separation technology used to extract sand from oil or gas from crude oil in the North Sea and arguing that that the technology behind Dyson's dual cyclone was, nothing not generally already known within the industry. The triple-vortex vacuum cleaner was, however, banned from sale in 2000 after the company was found guilty of patent infringement. The judge considered that the whirlwind principle remained the same and that "throughout the development of the Vortex cleaner Hoover was aware of patent claims concerning cyclonic vacuum cleaners, including those claims contained within Dyson patents." Hoover was ordered to pay damages to Dyson and forced to launch a new machine, which did not

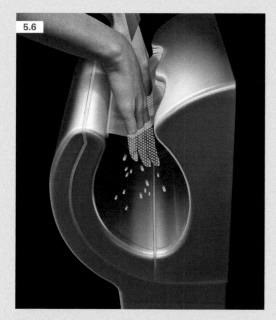

Figure 5.6
Dyson Airblade. Dyson's hand dryer uses a small brushless
motor, which runs at 100,000 revolutions per minute to force
filtered air through two apertures, creating fast-moving
sheets of air that "scrape" water from wet hands. It's quick,
hygienic, and uses significantly less energy than traditional
warm-air hand dryers.
Photograph: Dyson Ltd.

infringe the patent. Dyson, however, lost the
last round of the High Court battle by failing to
stop Hoover using the "Vortex" trademark on
its bagless cleaners.

Dyson has continued to develop new
products such as vacuum cleaners, washing
machines, fans, and hand dryers. They all
incorporate degrees of innovation through new
technologies of user-led insights, and all are
enabled by the protection afforded by intellectual property rights including around 1,900
patents worldwide.

Figure 5.7
DC59 cleaner. This clever product recognizes that many people no longer want to spend a long time dragging a heavy vacuum
cleaner around a house and provides a lightweight alternative using battery power that works with shorter period of cleaning.

Sales

The Industrial Revolution meant that more and more products could be made using mass-produced and -powered heavy machinery and transported with powered machines to consumers over much wider geographic territories. Marketing developed initially through a need to provide product information to these new and distant markets. As more and more companies have followed this expansion providing more and more choice, the role of marketing has expanded further still to include the selling techniques needed in the face of intense, global competition. This final section outlines some of these sales activities as well as the marketing trends that are relating directly to process of product design.

5.8

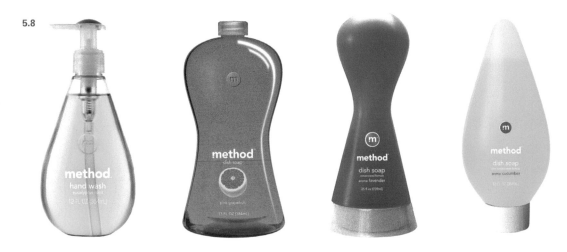

Figure 5.8
Hand wash and dish soap packaging by Karim Rashid for Method products. Method raised the question as to why many people hide their bathroom products in a cupboard. Packaging design is, therefore, one strong element of their marketing mix. Karim Rashid's packaging designs include teardrops, truncated cones, and a squeeze bottle design that includes an unusual pinch-neck feature. These elements combine to make a more visually interesting and tactile experience for consumers. Rashid has designed a vast array of products, and over 2,500 have been put into production.

Marketing mix

The marketing mix includes the four P's of Product, Price, Place, and Promotion, which are regarded as the sales tools that can be manipulated to enable a product to be sold in the most effective way (although it's often noted that attention to People, Processes, and Performance should also be in the mix). A designer should have a key role here because they have acute insights into the mindset of the market, about what consumers like, what they might pay, and how they behave and they should hence be involved in creating the optimum product marketing mix.

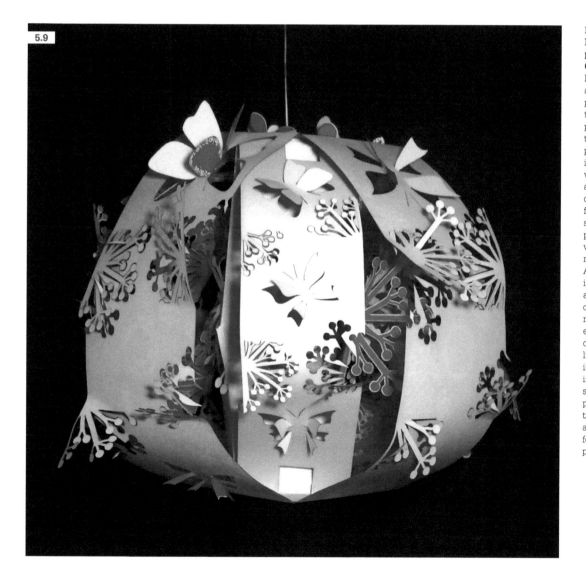

5.9

**Figure 5.9
Light bulb
packaging by
Olivia Cheung.**
Packaging has
always had a
protective role in
the transport of
products, but in
this vastly com-
petitive world the
instantaneous
view and appear-
ance of a product
can have a pro-
found effect on its
sales. Therefore,
packaging forms a
vital part of pro-
moting a product.
Although packag-
ing is important, it
also attracts criti-
cism as a cause of
much waste and
environmental
degradation. This
light bulb packag-
ing can be turned
into a filigree light
shade once its
protective func-
tion is complete
and aims, there-
fore, to reduce
packaging waste.

Distribution and retail

E-business has transformed the way that many people research and buy products, providing suppliers with an opportunity to reach markets quickly with reduced retailing costs. Designers might here be concerned with product issues of transport weight and protection, and assembly and operation instruction. It is not, however, the end of store-based retailing. Ikea, for example, has become the leading retailer across the globe in part by creating stores that are in themselves places of destination with unusual cafes. Ikea offers simple products that appeal to the target demographic at lower prices to take the sting out of paying for traditionally expensive furniture.

This philosophy of a virtual business cycle—high volumes leading to lower costs, leading to lower prices, leading to more sales—is seen in both e-business like Amazon and traditional companies such as Ikea and Walmart. Both require effective distribution and background systems and processes to make them work. Good design provides the foundation for each model, however, and illustrates the way in which design intent and business strategy go hand in hand.

Branding

Marketing has traditionally been involved with informing and persuading consumers to buy products sales based on the promoting of its functions and performance. Branding represents a move in the late twentieth century toward sales based less on product features and more on a product's emotive attributes. For example, consider the design of a new vacuum cleaner that has a bigger motor than its rivals. Sales might be based on the merits of owning a cleaner that has more power, but how much does this really excite consumers? Perhaps instead, they would be more interested in a vacuum cleaner that cleans so well that allergens within the house are reduced so that (tacitly) the consumer may live longer—this is a much more powerful message! Through branding, the affirmative and semantic features a designer may have included in the design can link to the core values of a consumer. Where these links are enhanced through lifestyle emphasis, celebrity endorsement, and imagery, they become extremely powerful selling techniques.

THE POWER OF BRANDS

There are debates that suggest the power and intrusiveness of brands is not healthy. It is also argued that innovation can decline when the designer's role is simply to produce an emotionally rich, brand-centric product to the detriment of the performance, functions, and features of a product.

"Establish contact with the subconscious of the consumer below the word level. They work with visual symbols instead of words . . . they communicate faster. They are more direct. There is no work, no mental effort."
Rosser Reeves

5.10

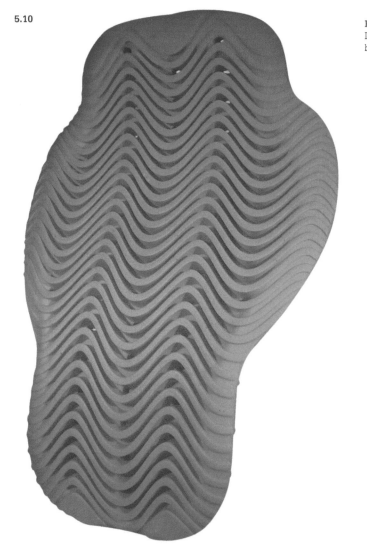

Figure 5.10
D30® Smart Skin in regularly featured brand orange.

Mass customization

Chapter 1 of this book described how designers try to understand user requirements and interpret these into a uniformly desirable product and included a section on co-creation as a method for reducing design errors by allowing users to directly create the product they want for themselves.

The intention of **mass customization** is similar to that of co-creation in enabling consumers to secure their own, personal design, but it aims to do this by allowing a mass of people to customize a product to their own design requirements rather than design it themselves from the outset. This is similar in principle to the way that individual products used to be crafted prior to industrialization, but the difference now is that modern technology is used to refine a product by drawing together many of the areas identified throughout this book: individuality, customization, co-creation,

Figure 5.11
Dahlia by Janne Kyttänen for Freedom of Creation. Danish designer Janne Kyttänen's light is made from laser-sintered polyamide and, as the product's name suggests, was inspired by the mathematics and layout of nature's dahlia flower. Organizations such as Freedom of Creation are expert in the production of bespoke designs using rapid prototyping.

5.11

rapid prototyping, rapid manufacture, modular design, just in time, and knowledge management.

For example, a consumer might go online and select the features and dimensions they personally require for a product. If this is clothing or a wearable product, they might scan and supply their own personal measurements. A knowledge management system might then generate a CAD model based on a parameterized and modularized product architecture. The unique product might then be produced using rapid manufacturing techniques and shipped out via globalized supply chains.

Buy back

To make a product so durable that it lasts for longer and longer periods might sound like commercial suicide if the company ends up selling fewer and fewer products. If, however, the market recognizes this durability, then the reputation of the company is enhanced, and sales can increase through greater consumer loyalty as well as from new customers. This partnership mode of business is being increasingly driven by the environmental agenda as more and more people recognize the need to create more long-lasting products as part of a more sustainable future. It

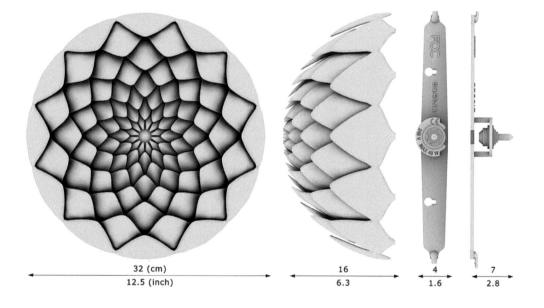

32 (cm)
12.5 (inch)

16
6.3

4
1.6

7
2.8

"It's actually very easy to design and manufacture a toaster that will last 20 years; that can be done. What's not so easy is to design and manufacture a toaster that someone will want to keep for 20 years."
Jonathan Chapman

5.12

**Figure 5.12
Mirra Chair by
Herman Miller,
Inc.** The Mirra
Chair is the first of
Herman Miller's
product range
to be designed
using a life-cycle
methodology and
has stemmed from
the company's
commitment
to becoming a
sustainable busi-
ness. The chair
is designed for
longevity; it has
a twelve-year
warranty period
and is 96 percent
recyclable without
any compromise
in performance or
style.

is the polar opposite to the concept of
disposable products or those with planned
obsolescence,and is referred to through a
number of other terms and systems such as
the *circular economy* or *regenerative mindsets*.
Some furniture companies in particular have
embraced this concept, but more industries
may follow through business choice or through
government regulatory requirement.

More enduring products usually require
a change in corporate strategic thinking,
including for example agreements to buy back
or repair products that break in service. These
can have significant design implications.
Products need to be robust and easy to
disassemble and repair with an increased
expected lifetime. It's not just about longevity
in performance, however, as the product should
retain its appeal over a long time span, and
the designer should be aiming for timeless
products.

"For me, the concept of design is more than object-oriented; it
encompasses the design of processes, systems and institutions
as well. Increasingly, we need to think about designing the
types of institutions we need to get things done in this rapidly
accelerating world."

John Seely Brown, Chief Scientist, Xerox Corporation

Product service systems

A product service system extends even further the notion of a working partnership between consumer and business. In this case, the manufacturer removes the product sale completely and instead provides the function that the product would have performed as a loan service. The most apparent example of this is in photocopying where companies rent a copying machine rather than buy a copier. This is a win-win-win situation. The customer gets what they want, which is reliable, photocopied paperwork which they pay for on a page by page basis. The manufacturer wins because it gets a continuing income stream through the loan and service agreements. Notably the environment also wins because the onus is on the designer to create a long-lasting and reliable product as it is the manufacturer, not the company, which will have to repair it if it breaks.

The trend for renting rather than purchasing includes products such as skis, car shares, bike rentals, and numerous digital applications replacing tangible products such as maps, video rental, or compasses. Product service systems may well become a much bigger factor for design in the future and may see a return to the rental of products (e.g., refrigerators and televisions) that was prevalent in postwar years. The implications for designers here might include designing products that are more robust, or that are more modular allowing for upgrades and improvements.

5.13

Figure 5.13
Reduced Carbon Footprint Souvenirs by Hector Serrano. Souvenirs and novelty gifts might be small, but they can have a high carbon footprint, especially if they are made in one country, transported overseas, then stored and redistributed in another country. The digital files for these gifts, however, can be e-mailed to the recipient and rapid prototyped locally, thus incurring a much lower energy requirement. This concept idea works on the assumption that rapid prototyping technology is becoming ever more available.

Case study
Vertu

The Vertu Company was originally established in 1998 by Finnish company Nokia to produce mobile phones with an emphasis on phones as fashion accessories. The strategy to achieve this has been to develop a luxury-end, upmarket brand profile, doing for phones what Rolex has done with the watch market.

Pioneering any new market can be challenge, but the high end market can be considered particularly bold for many reasons; A troubled global economy tends not to favor high-value goods and the company's position stands in direct contrast to the move seen in many industries towards more economical products. The brand is not supported by years of trading, which could bring it the heritage of, say, Rolex or Rolls-Royce. Mobile phones also represent rapidly moving technology, so the target market must be persuaded to spend a lot of money on a product that performs in much the same way as cheaper models do and which might become dated very quickly. Finally, the high-class market is highly competitive, and rivals might include not just other upmarket phone companies but also other luxury brands such as Louis Vuitton, Prada, or Omega who may launch mobile phones as part of their own brand extension strategies. Good design has, therefore, been the key to overcoming these issues and making the corporate strategy successful,

and the design team at Vertu has been headed up by principal designer Frank Nuovo who has achieved widespread recognition across the United States and the world for his industrial design work. Under his leadership, the team has sought to achieve this success by understanding and providing exactly what luxury market users require: top of the range quality, individual craftsmanship, exclusivity, and individuality. Vertu phones, therefore, use luxury materials such as white gold, platinum, and premium leather. Sapphires, second only to diamond in gemstone hardness, are used to create scratchproof screens and rubies are used for low-friction bearings on many of the product's moving parts. There is an emphasis on ensuring that mobile phones "connect" with their market—an empathetic rather than purely technological approach. Phones include a patented concierge key that links customers directly to a round-the-clock service team offering lifestyle support, in much the same way as a hotel concierge operates. There is also an understanding of the way that its target consumers appreciate craft rather than mass production. Each telephone is constructed using handmade techniques, and in the case of Vertu's Signature range, this incorporates 288 components and 96 quality control inspections. This time-consuming but artisan approach represents a move toward custom-made and crafted products. To further enhance design and the functional attributes of exclusivity and individuality, Vertu's products are sold only through its own branded stores or through specialized or high-end stores such as Harrods (London) and Barneys (New York).

In more recent years, the company has released a full range of mobile phone products that vary in emphasis on the different elements of these user requirements: materials, performance, craftsmanship, individuality, and uniqueness. The company has subsequently established a strong business, which has now been partially spun out by Nokia. To remain strong, however, the design team must continue to face up to new design challenges. For example, should new products be made modular, allowing technological upgrades to be easily made but at the expense of the company's custom-made philosophy? Should the company continue to use its own or bought in technologies? What will future phones look like and how will they work—will they morph more toward wearable products in a similar way to watches or jewelry. As before, the answer to these questions lies in the designers at Vertu to continue to understand how to make their brand meet the needs of their customers and their markets.

Figure 5.14
Signature by Vertu. Vertu's Signature range incorporates some of the most expensive and exclusive mobile phones in the world. Vertu is a crafts-based manufacturer, and the company's attention to materials makes products that can be likened to works of art.
Photograph: Vertu

Chapter summary

The final chapter in this book has set out to explore some of the postproduction issues for new products, how these relate to designers and particular why these issues need to be considered at an early stage in the design process rather than once a product has left the factory gate.

The first section looked at issues of testing and protecting products. When products get to market, they need to work and its best to know through testing before release how the product actually performs. As well as financial penalties, there can be severe legal penalties if they do not. A designer must, therefore, work systematically to ensure that they have considered all possible risks involved and reduced these to the minimum level required to make a product safe. Intellectual property rights are available to help protect products from being copied once they are launched. It's important to remember that there are rigid rules about obtaining protection that must be followed, and that an idea by itself is rarely capable of being protected—it needs translating into features and designs.

The final section looked at the role of marketing and branding in creating commercial success. The selling of products through the power of emotion might take the product back to the core values a designer explored at the very outset of a project. The use of mass customization, buy-back, and product service systems provided illustrations to some of the newer marketing and business techniques, which may be how products of the future will all need to be sold—and designed accordingly. In a product service system, the designer has to consider the creation of a robust product and understand completely what it is that the customer really needs. This might then take us right back again to the beginning of this book.

Assignments

1. Conduct a risk assessment on a product of your own choosing, and incorporate design features that reduce the hazards and chances of user injury.

2. IPR can be a contentious subject. Some might argue that it is simply wrong to own and protect things, particularly, for example, in the field of medical or food products. IPR can also encourage monopoly positions that are an anathema to free trade. On the other hand, IPR can be said to encourage investment and research into the types of new ideas that benefit us all. Research both side of this argument; then present your opinions on this emotive subject.

3. Look at the products and services being offered by a range of brand conscious companies. Dig beneath the imagery, identify the brand values, and evaluate them.

4. Select a product at random. Define the purpose of the product with one key function, and consider how you might provide this function as a service rather than a product.

5. Using the criteria of being the simplest but having the biggest impact on people or society, which new product in the past one hundred years do you think has been the most innovative?

Conclusion

This book has aimed to illustrate that the process of creating well-designed products starts and ends with people.

As was stated at the beginning, the sequential step-by-step format laid out in the book is by no means the definitive model of the process and is rather idealized. The reality is that it is unlikely to be very linear. Some projects skip parts of the process, some take more time and effort in places than others, sections pop up in unexpected places with more—or less—frequency than might be ideal, some parts take place concurrently and some parts go back and forth so that the whole process can ebb and flow and sometimes end up going nowhere at all.

"My job is to make our products obsolete before our competitors do."

Akito Morito, Chairman, Sony

It does, however, hopefully highlight just how rich the process is and may have surprised anyone who thought design was just about ideas, CAD, or sketching. Other topics raised have included anthropology, legislation, experimental analysis, research methods, empathy, decision making, attitude, quality, and even personal philosophy. The list of skills, knowledge, and capabilities a designer needs might seem endless. In fact, there are debates within academia as to whether product design is just too big to cover in one course and should be split into smaller courses that specialize in parts of the process—design research, user experience, or engineering designer—mirroring some of the practices seen in industry.

It would be rare to find one designer who is brilliant at all of these things, or indeed who needs to be brilliant at all of these things. Different designers may have particular strengths in different areas, which may be down to genes, or training, and sometimes down to the type of occupation, and the demands of an individual designer craftsman may be very different to a designer of components in a large multinational company or where you might find design teams.

This variety and depth of know-how does, however, mean that designers might always be drawing on the experiences and lessons they have learned in life in more ways perhaps than any other profession. It also means that they can always learn more, improving areas of strength and developing areas of weakness. For those with an opinion, a thirst to face challenges, and a desire to learn and develop, then there can be no better profession to choose than that of product design.

A great deal of knowledge is also simply generated by going through this design process: how best to research market needs, how to make products, how to work new materials or old materials in new ways, how to create functions through clever mechanisms, and how to please people. For individuals, this knowledge might simply be expertise retained in their own thoughts, captured through reflection and experience. This experience is highly valuable, and it might be surprising to learn that the global trade in knowledge is actually greater than the global trade in goods. For example, this might mean your knowledge in how to make a particular product is worth more than the income generated by the sales of the product.

For larger organizations, capturing knowledge across a vast range of designers working on a variety of projects is even more complex and requires more sophisticated knowledge management systems to ensure that expertise is not just captured but retained, valued, disseminated, and used across the whole company. This can require solid systems such as information systems, networks, and

Figure 6.1
Rubik's cube.
Designed originally to demonstrate 3D geometry in class, the Rubik's cube has gone on to be one of the best-selling products of all time, and is the bestselling toy product of all time.

6.1

database management, as well as attention to softer cultural systems, such as encouraging serendipity, supporting teamwork, or ensuring that staff members recognize the value in their work.

Some companies are better at this than others—and much better than others at releasing successful new products. Surveys suggest that the ability of many organizations to innovate is actually quite low. Sometimes it is obvious when the process has gone wrong through products that look or work badly. Normally, however, these products are quickly pulled from the market, and the real damage is usually hidden through lost costs from products that take too long to get to the market, products that allow better products from competitors to take a bigger market share, and those products that involve expensive redesign. Companies that are good at the design process—and at knowledge management—become better and better at generating new products and developing a strategic advantage by becoming associated synonymously with innovation. Managing the process and knowledge management are beyond the scope of this book, but subjects such as product lifecycle management (PLM) and concurrent engineering might lead toward further reading.

Ultimately, and pleasingly, however, the efficiency of the process does not define successful products nor does the experience of the designer (although these may increase the likelihood of success). The products themselves define success. This answer then begs the question how do you measure success? By sales? By production numbers? By profit? By awards won? Or perhaps by the benefit made to mankind?

There are few answers here. You can set your own rules, and it can be a surprise to some that design can be as much about philosophy and personal perspective as it is can be about determining the right size bolt or which CAD package to use. Sometimes these philosophical debates are polemic with fierce advocates on one side or the other, but more often the issues are cloudy. Seeing through the mist can be an important part of a designer's life, and through this clarity and his or her design work, the designer can have an influence on world affairs—sometimes grandly, sometimes in small steps. Such is the practice of design linked to the issues of life itself.

"If money is your hope for independence you will never have it. The only real security that a man will have in this world is a reserve of knowledge, experience, and ability."

Henry Ford

Glossary

aesthetic A term used to describe the look of a product but also the way that the product feels and interacts with all senses and emotions, calling on issues of perception and reflective critique.

alloy A substance containing two or more elements (usually metals). For example, an alloy of aluminum and copper is usually used in design rather than pure aluminum because it is much stronger. Designers should specify the precise alloy required rather than simply "aluminum" based on the precise alloy needed to meet the required material performance.

alpha testing Experimental in-house investigation of a developed product, prior to in house testing of the product in its working conditions (beta testing) or testing with a limited number of pilot consumers (gamma or field testing).

anthropometry A branch of ergonomics that looks at the physical properties of people and that draws on statistics in its application to the design of products.

benchmarking Studying and comparing products to establish best practice. It is a useful tool for uncovering prejudices that may exist in thinking.

bill of materials A list of subassemblies, components, and/or raw materials that make up a product, usually presented in a hierarchical format.

break even The point at which income from a product precisely covers the costs that have been made in development and production, after which profits may be expected. It is usually measured in units of time or production numbers

cognitive modeling A model of understanding based on psychological principles of how users perform tasks. It can be used to help the design of interfaces based on behavioral expectations. For example, if people are known to move away from wasps based on their visual and instinctive reaction, then the colors black and yellow might help to provide warnings for a dangerous product.

concept A largely unproven proposition that meets the broad objectives of a product idea. It can be a theoretical vision, a written description, or a modeled view.

concurrent engineering A systematic attempt to undertake the stages in new product development simultaneously rather than in a linear, step-by-step process. Perhaps like reading a page from each chapter of this book at the same time, before moving onto the next page of each chapter.

critical to function (CTF) The engineering tolerances that are typically different from normal standards and that are vital in the effective operation of a product or component. These may be expressed within a technical drawing or highlighted by specific text.

datum A point, line, or plane that is used as a theoretical or actual anchor point against which other dimensions are referenced. It is an essential aspect of engineering and CAD.

decision making In new product development process is the action of selecting or reaching a judgment on a range of options. The decision is often made at a key stage (gate or gateway) by comparing the products potential against a set of company specific criteria (screening). From the Latin term *to cut off* (as in the mental process of reducing options).

Design for X The principles, tools, and guidelines that aid in the process of Design for Manufacture (DFM), Design for Assembly (DFA), Design for the Environment (DFE), or Design for any Function X (DFX). Design to cost means treating cost as an independent parameter in the design process that must also be met.

ergonomics A science that studies the interface between people and products. It is not limited to size and anthropometric data and includes issues of safety, comfort, ease of use, aesthetics, and operability.

ethnography A qualitative, action-based research method for understanding customer needs based on, and using tools of, anthropometry and culture.

experience design An assessment of the experiences of product users drawing on a wide range of sources including physiological, sociological, psychological, and ideological subject areas. It is closely related to how products emotionally affect the user (affective design).

failure mode and effect analysis (FMEA) A systematic assessment of the ways in which a product or a design might fail with suggestions and methods to reduce the likelihood of this happening. Designing out errors is referred to as *mistake proofing*, or *Poka-Yoke* from the equivalent Japanese term.

failure rate A term that means either the number of a specific product failing at the manufacturing stage or the numbers of products released to the market that fail to meet customer expectations.

finite element analysis (FEA) A method of calculating stress in a product that allows it to be designed with optimum material dimensions and reduced dangers of breaking. Computer software breaks a component geometry into elements and links a series of equations to each, which are then solved simultaneously to evaluate the behavior of the whole. Used for solving a variety of engineering problems including structural analysis and thermal and fluid dynamics and can be linked powerfully to CAD drawings.

focus groups Meetings with product stakeholders (e.g., users, buyers, producers) to explore product needs or to assess product ideas. Groupthink is a danger at these sessions and is a symptom of individuals not expressing personal beliefs in favor of a common group consensus.

inclusive design Design that seeks to meet the needs of all users (including, for example, the less abled). It is synonymous with *universal design*.

innovation The transformation of an idea into a novel, saleable product, operational process, or new service (from British Standard BS 7000).

intellectual property Design know-how and its manifestations, protected by statutory and legal articles referred to as the intellectual property rights.

International Organization for Standardization (ISO) An agency that coordinates and publishes standards between countries. For example, ISO 9000 is the set of guidelines that an organization should adopt to ensure quality management and assurance is to a satisfactory standard, and ISO 9001 relates this to new products.

knowledge management A process of managing the capture, organization, and dissemination of knowledge. It is particularly relevant to new product development that can use vast quantities of data.

lean manufacture The principle of linking consumer demand to production so that products are made only when required. It works through methods such as just in time, modular design, and robust design and makes financial savings by reducing raw materials, components, and stock are minimized. It can also facilitate innovation by making it easier to introduce new products.

life cycle assessment (LCA) This term has come to mean the systematic review of a product's environmental impact over its entire life, from raw materials to disposal.

mass customization A principle of linking an individual consumer's requirements to create a unique product design that is constructed using mass production techniques.

modular design The ability to realize a range of unique products using building blocks of basic parts or modules. A platform product describes the derivatives from which a product family can be developed.

new product development (NPD) A term used to describe all of the activities that generate new products including the marketing, design, research, manufacturing processes, and supporting business functions. Synonymous with *total design*, a multidisciplinary process that takes a product through a series of stages from market need through to sale and disposal.

parametric design A feature of CAD that allows users to define links, dimensions, and constraints within a drawn design. Individual changes to the drawings are then automatically reflected throughout the design.

part classification A way of defining products according to their attributes such as geometry or symmetry. Where product numbers and components are large, this is a way of controlling the design of parts, avoiding waste and repeated effort.

process capability A statistical tool that defines the repeatability and consistency of a manufacturing process. It relates to a product's design parameters and machine capabilities.

product brief The encapsulation of the requirements of a product. The brief helps to define the scope of a product (i.e. the sum of the market needs that are being tackled by this particular product) and can be textual or visual in nature, broad or specific. A fully parameterized description including quantified criteria and design targets is usually referred to as a specification or PDS (Product Design Specification).

product life cycle This term can be used to define the stage at which a set of products are performing in the sales environment (introductory, growth, maturity, or declination stages) or by the stage a particular product is within the development process (idea, concept, development, production, sales, and disposal stages).

prototype A working model built for testing and evaluation. Models and prototypes can range in quality from rough manifestations to full facsimiles of production units. A proof of principle prototype (POP) may evaluate a specific product function.

rapid prototyping (RP) The form of prototype produced by slicing a CAD model data into layers and reproducing the layers in a physical form using a variety of techniques such as stereolithography or selective layer sintering. Using the same techniques to produce a series of actual parts is referred to as rapid manufacturing, and to enable production tooling is referred to as rapid tooling.

rendering A two-dimensional image of a concept that can be hand-sketched or produced by CAD or specialist visual software.

reverse engineering The process of scanning a product to digitally capture its geometry, which can then be reproduced as a CAD drawing. It also means to learn more about a design by dismantling an existing product.

robust design A product designed in a way that limits the product to variation through misuse and increases the probability that it will perform as intended.

six sigma A statistical tool built around five interconnected phases: define, measure, analyze, design, and verify. The tool uses methods such as QFD and FMEA in an attempt to ensure that the customer's needs are met by large-scale manufacturing methods with only very limited product failures.

standard cost The predetermined or planned cost of manufacturing a single product (unit).

supply chain The trail of interconnected subcontractors and retailers who supply raw materials plus the components and parts that are outsourced. Complex global logistics and scheduling require this element of an organization to be managed.

Taguchi method An approach to quality developed by Genichi Taguchi that aims to limit product failures by reducing their sensitivity to the factors that cause products to fail. This quality approach uses statistical and experimental methods to achieve this.

tolerance The specified upper- and lower-dimensional limits of a component in which it must be manufactured to function correctly.

Further resources

Resources for design are as wide ranging as the subject area itself, but you might try some of the following examples as places to obtain further information or just as sources of inspiration.

Websites
- www.coolhunting.com
- www.core77.com
- www.delphion.com
- www.designboom.com
- www.designmuseum.org
- www.designnation.co.uk
- www.dezeen.com/design
- www.howstuffworks.com
- www.ted.com
- www.trendhunter.com
- www.tuvie.com
- www.yankodesign.com

Places to visit
- Design Museum, London, United Kingdom
- Exploratorium, San Francisco, California, USA
- International Design Centre , Nagoya, Japan
- National Design Museum, Cooper-Hewitt, Smithsonian Institution, New York, New York, USA
- Red dot Design Museum, Essen, Germany (also in Singapore)
- Vitra Design Museum, Weil am Rhein, Germany
- Victoria & Albert Museum, London, United Kingdom

Exhibitions
- 100% Design, London, United Kingdom
- Decoration + Design, Australia
- Designers Block, Global
- International Furnishing show, Cologne, Germany
- London Design Festival, London, United Kingdom
- Milan Furniture Fair, Italy

Competitions and awards
- Audi Design Foundation
- Design & Art Direction (D&AD)
- Product of the year
- Design Sense Award
- Electrolux Design Lab

Journals
- *Design Studies*
- *Design Week*
- *Domus*
- *ICON*
- *International Journal of Design*
- *Science*
- *Technovation*

Further help
- Design Council
- Design Research Society
- European Academy of Desig
- IDEO HCD toolkit
- Institution of Engineering Designers
- Royal Society for the Encouragement of Arts, Manufacture, and Commerce

Further reading

Research
- Clarke, A. (2010) *Design Anthropology*. Springer.
- Papanek, V. (1982) *Design for the Real World*. Thames & Hudson.
- Rawsthorn, A. (2013) *Hello World: Where Design Meets Life*. Hamish Hamilton.
- Robson, C. (2002) *Real World Research*, 2nd edition. Blackwell Publishing.

Concept development
- Bramstone, D. (2009). *Idea Searching*. AVA Publishing.
- De Bono, E. (2009). *Think! Before It's Too Late*. Random House.
- Eissen, K., and Steur, R. (2012). *Sketching—The Basics*. BIS Publishers.
- Hallgrimsson, B. (2012). *Prototyping and Modelmaking for Product Design*. Laurence King Publishers.
- Harman, J. (2013). *The Shark's Paintbrush*. Nichols Brealey Publishing.
- te Duits, T. (ed.) (2003). *The Origin of Things; Sketches, Models, Prototypes*. Fortis.

Product development (engineering)
- Ashby, M., and Johnson, K. (2010). *Materials and Design*, 2nd edition. Butterworth Heinemann.
- Davey, A. (2003). *Details: Exceptional Japanese Product Design*. Laurence King.
- McLellan, T. (2013). *Things Come Apart*. Thames & Hudson.
- Moaveni, S. (2010). *Engineering Fundamentals: An Introduction to Engineering*. Cengage Learning.
- Thompson, R. (2007). *Manufacturing Processes for Design Professionals*. Thames & Hudson.
- Voland, G. (2003). *Engineering by Design*, 2nd edition. Addison Wesley.

Product development (human factors)
- Chapman, J. (2005). *Emotionally Durable Design*. Routledge.
- IDSA. (2003). *Design Secrets: Products*. Rockport.
- Fukosawa, N. (ed) (2009) *Naoto Fukosawa*. Phaidon.
- Norman, D. (1998). *The Design of Everyday Things*. MIT Press.
- Martin, B., and Hanington, H. (2012). *Universal Methods of Design*. Rockport.
- Pheasant S. (2001). *Bodyspave*. Taylor & Francis.
- Rubin J. (2008). *Handbook of Useability Testing: How to Plan, Design and Conduct Effective Tests*, 2nd edition. Wiley.

Production
- Atrill, P. (2011). *Prentice Accounting and Finance for Non-specialists*, 7th edition. Hall.
- Boothroyd, G., Dewhurst, P., and Knight, W. (2010). *Product Design for Manufacture and Assembly*. CRC Press.
- Drury, C. (1998) *Costing: An introduction*, 4th edition. International Thomson Business.
- McDonald, J., and Ryall, J. (2001) *Rapid Prototyping Casebook*. Professional Engineering Publishing.
- Simons, C., Maguire, D., and Phelps, N. (2009). *Manual of Engineering Drawing: Technical Product Specification and Documentation to British and International Standards,* 3rd edition. Newnes.

Product launch
- Baines, P. (2011). *Marketing*, 2nd edition. Oxford University Press.
- Bainbridge, D. (2012). *Intellectual Property*, 9th edition. Longman.
- Howard, A. (1997). *Safer by Design*, 2nd Edition. Gower.
- Miller, C. (2004). *Product Liability and Safety Encyclopaedia*, 2nd edition. Oxford.

.

Index

Acknowledgments and credits

With grateful thanks to Kate Duffy, Brendan O'Connell, Felicity Cummins, Christopher Black, colleagues at the University of Brighton, and all the other kind people who have contributed to the book. Special thanks to Sue, Barney, and Seb.

Photo credits

Chapter 1

Figure 0.1
Mirra chair by Herman Miller, Inc.

Figure 1.1
Morph device concepts by Nokia

Figure 1.2
Personal environment monitor by Lapka™

Figure 1.3
Urban beehive by Philips

Figure 1.4
Merck Serono easypod® design by PDD Group Ltd.

Figure 1.5
Switch screwdriver by Mr. Yunlong Liu, Mr. Peng Jia, Mr. Peng Cheng, Mr. Dongdong Wang, and Miss Yaoyao Xin

Figure 1.6
Sun Jar by Tobias Wong for Suck UK

Figure 1.7
Nike Air Jordan XX3

Figure 1.8
Nike Flyknit trainers

Figure 1.9
Neatly nesting mobile phones by Kyle Bean

Figure 1.10
Witches kitchen cookware

Figures 1.11
Brunton Hydrogen Reactor™

Figure 1.12
LotusSports Pursuit Bicycle by Mike Burrows

Figure 1.13
Smart Car

Figure 1.14
Philips food printer concept

Figure 1.15a
iPhone by Jonathan Ive and the Apple design team

Figure 1.15b
Photo by Gavin Roberts/T3 Magazine via Getty Images

Figure 1.16
BeoLab loud speakers by Bang & Olufsen

Figure 1.17
Eclosion by Olivier Grégoire

Figure 1.18
Leatherman® Skeletool

Figure 1.19
Smart Crossblade

Figure 1.20
Plume Fountain Pen by Vivien Muller

Figure 1.21
Nissan IDx NISMO®

Figure 1.22
Seed Cathedral, UK Pavilion at Shanghai Expo 2010

Figure 1.23
Spun Chair

Figure 1.24
Olympic Cauldron

Chapter 2

Figure 2.1
Maglev concept washing machine by Jakub Lekes

Figure 2.2
Hyper Fast Vase by Cedric Ragot

Figure 2.3
Teiko by Anthony Dickens

Figure 2.4
Glass Tap by Arnout Visser for Droog

Figure 2.5
Not a lamp by David Graas

Figure 2.6
Mezzadro Stool by Castiglioni

Figure 2.7
Opto reading glasses by Xindao B.V.

Figure 2.8
Graphene aerogel

Figure 2.9
Segway by Dean Kamen

Figure 2.10
Massive Infection by Pieke Bergmans

Figure 3.23
First by Michele De Lucchi for
Memphis

Figure 3.24
Super by Martine Bedin for
Memphis

Figure 3.25
The iMac by Apple

Figure 3.26
OXO Good Grips utility knife by
Smart Design

Figure 3.27
Zippo Lighter

Figure 3.28
Antique fridge by Meneghini

Figure 3.29
Sandbug

Figure 3.30
Gofer screwdriver

Chapter 4

Figure 4.1
Copper shade pendant light by
Tom Dixon

Figure 4.2
Deena Low table for Habitat

Figure 4.3
Levitator floating speakers by
James Coleman

Figure 4.4
Segway development

Figure 4.5
Achim Menges Morpho Design
Experiment No. 1

Figure 4.6
Napshell working model

Figure 4.7
Hidden.mgx vase by Dan Yeffet
"Jelly Lab" for Materialise

Figure 4.8
Mini

Figure 4.9
Quin.mgx lamp by Bathsheba
Grossman

Figure 4.10
Bathtub couch by Max McMurdo
for Reestore

Figure 4.11
Omi.mgx lamp

Figure 4.12
Femur stool

Figure 4.13
Spun chair by Thomas
Heatherwick

Figure 4.14
Wire chair by Tom Dixon

Figure 4.15
Little Tikes Cozy Coupe® Car

Figure 4.16
Zanzibar Tower ZTR and Hot
Spring radiators by Bisque

Figure 4.17
AI Stool.mgx for Materlise.mgx by
Assa Ashuach

Figure 4.18
Freedom of Creation 610 Lamp by
Jiri Evenhuis and Janne Kyttänen

Figure 4.19
Slab Chair

Figure 4.20
Fresh Fat Easy Chair

Figure 4.21
Door handle by Dieter Rams

Figure 4.22
JCB teletruck

Figure 4.23
Quinny Buzz pushchair

Figure 4.24
POÄNG chair by Noburu
Nakamura for Ikea

Figure 4.25
Kitchen scales 1007 model by
Salter Housewares

Chapter 5

Figure 5.1
Dyson DC54 cleaner

Figure 5.2
EvoShave by EvoShave Ltd.

Figure 5.3
Around Clock by Anthony
Dickens for Lexon

Figure 5.4
Dyson technology

Figure 5.5
Blue Touch by Royal Philips N.V.

Figure 5.6
Dyson Airblade

Figure 5.7
DC59 cleaner

Figure 5.8
Hand wash and dish soap packaging by Karim Rashid for Method
products

Figure 5.9
Light bulb packaging by Olivia
Cheung

Figure 5.10
D30- Smart Skin in regularly
featured brand orange.

Figure 5.11
Dahlia by Janne Kyttänen for
Freedom of Creation

Figure 5.12
Mirra Chair by Herman Miller,
Inc.

Figure 5.13
Reduced Carbon Footprint
Souvenirs by Hector Serrano

Figure 5.14
Signature by Vertu

Conclusion

Figure 6.1
Rubik's Cube